my revision notes

OCR A-level

BIOLOGY A

Frank Sochacki

Hachette UK's policy is to use papers that are natural, renewable and recyclable products and made from wood grown in sustainable forests. The logging and manufacturing processes are expected to conform to the environmental regulations of the country of origin.

Orders: please contact Bookpoint Ltd, 130 Park Drive, Milton Park, Abingdon, Oxon OX14 4SE. Telephone: (44) 01235 827720. Fax: (44) 01235 400454. Email education@bookpoint.co.uk Lines are open from 9 a.m. to 5 p.m., Monday to Saturday, with a 24-hour message answering service. You can also order through our website: www.hoddereducation.co.uk

ISBN: 978 1 4718 4226 9

First published in 2016 by

Hodder Education,

An Hachette UK Company

Carmelite House

50 Victoria Embankment

London EC4Y 0DZ

www.hoddereducation.co.uk

Impression number 10 9 8 7 6 5 4 3 2 1
Year 2020 2019 2018 2017 2016

Cover photo reproduced by permission of

Typeset in Bembo Std Regular, 11/13 pts. by Aptara, Inc.

Printed in Spain

A catalogue record for this title is available from the British Library.

Get the most from this book

Everyone has to decide his or her own revision strategy, but it is essential to review your work, learn it and test your understanding. These Revision Notes will help you to do that in a planned way, topic by topic. Use this book as the cornerstone of your revision and don't hesitate to write in it — personalise your notes and check your progress by ticking off each section as you revise.

Tick to track your progress 

Use the revision planner on pages 4 and 5 to plan your revision, topic by topic. Tick each box when you have:

- revised and understood a topic
- tested yourself
- practised the exam questions and gone online to check your answers and complete the quick quizzes

You can also keep track of your revision by ticking off each topic heading in the book. You may find it helpful to add your own notes as you work through each topic.

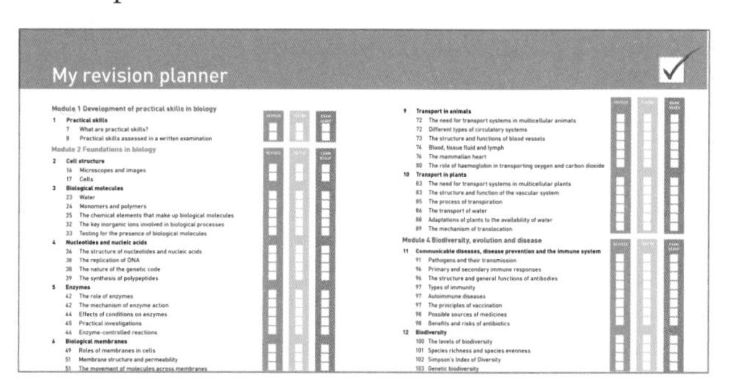

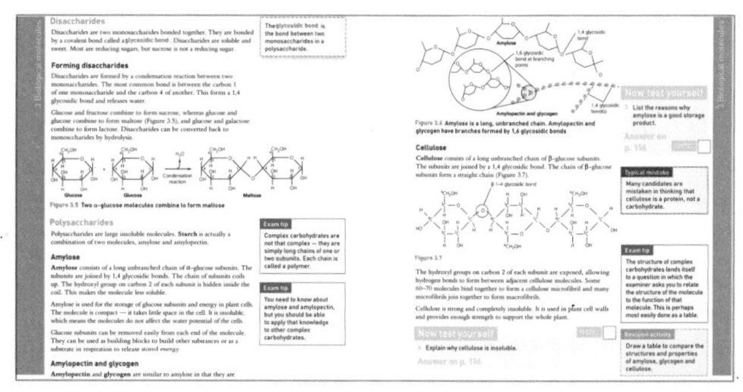

Features to help you succeed

Exam tips

Expert tips are given throughout the book to help you polish your exam technique in order to maximise your chances in the exam.

Typical mistakes

The author identifies the typical mistakes candidates make and explains how you can avoid them.

Now test yourself

These short, knowledge-based questions provide the first step in testing your learning. Answers are at the back of the book.

Definitions and key words

Clear, concise definitions of essential key terms are provided where they first appear.

Key words from the specification are highlighted in bold throughout the book.

Revision activities

These activities will help you to understand each topic in an interactive way.

Exam practice

Practice exam questions are provided for each topic. Use them to consolidate your revision and practise your exam skills.

Summaries

The summaries provide a quick-check bullet list for each topic.

Online

Go online to check your answers to the exam questions and try out the extra quick quizzes at **www.hoddereducation.co.uk/myrevisionnotes**

My revision planner

Module 1 Development of practical skills in biology

Module 2 Foundations in biology

Module 3 Exchange and transport

1 Practical skills

Practical skills assessed in a written examination

Your ability to apply your knowledge to practical situations will be tested. This includes questions that test your skills in:

- planning
- implementing
- analysis
- evaluation

Planning

REVISED

1 You may need to solve a problem set in a practical context.
2 You may be asked to make a prediction from a given hypothesis, or you may be given a prediction and asked how you could test it.
3 You should be able to consider the prediction and outline an experiment or investigation that will test the prediction. For example:

If you are given the prediction that yeast will stop respiring at 45°C, how would you test it?

4 You would need to describe how you would determine when respiration was taking place and how you would determine when it stops. You should then describe a method in which you tested yeast at a range of temperatures around 45°C to determine at what temperature it actually stops respiring.

Suitable apparatus, equipment and techniques

1 Select apparatus from a range of normal glassware and chemical reagents.
2 As respiration in all organisms releases carbon dioxide, it would be sensible to arrange a test that demonstrates the presence of carbon dioxide such as using lime water or bicarbonate indicator. You would then place a sample of yeast at different temperatures and demonstrate that it produces carbon dioxide at some temperatures but not at others.
3 You need to select apparatus that enables you to change the temperature of the yeast sample and pass the gas produced through the selected reagent.
4 Draw a diagram of the apparatus and how it is set up.

Application of scientific knowledge

1 Your plan should be based on suitable scientific knowledge. This knowledge should be applied to the practical context you are planning.
2 You will need to show that you understand the process of respiration well enough to know that sugars are needed and carbon dioxide is released. The process involves enzymes and it is these that are affected by the change in temperature.

Identification of variables

1 The independent variable is the variable you choose to change during the experiment. Temperature is the independent variable in this example.
2 The dependent variable is the variable that changes as a result of changing the independent variable. Production of carbon dioxide is the dependent variable in this example.
3 Controlled variables are any other factors that might affect the result, such as pH, volumes of solutions used, concentrations of solutions used.

Evaluation of the method

Show that you have considered whether the method is appropriate to test the desired hypothesis. Ask questions about the method.

1 Would your proposed method do the job?
2 Would it provide an answer to the question?
3 What data will you collect?
4 What will you do with the data collected?
5 Are there any limitations?
6 What level of uncertainty is there in the data collected?

In this example, you might ask:

1 Does it detect whether carbon dioxide is given off?
2 Can I alter the temperature or keep it constant?
3 Can the yeast respire if temperature is not limiting?

Your proposed method may not be perfect or provide a full answer. This is fine as long as you recognise this and say so in your evaluation.

Implementing

REVISED

As part of the written examination, you may be tested on:

1 how to use practical apparatus and techniques correctly
2 appropriate units for measurements
3 presenting your observations and data in an appropriate format

Presentation of results

Your results need to be presented in a table in the appropriate format:

1 There should be an informative title.
2 All raw data should be placed in a single table with ruled lines and a border.
3 Input or independent variable (IV) should be in the first column.
4 Output or dependent variable (DV) should be in columns to the right.
5 Processed data (e.g. means, rates, standard deviations) should be in columns to the far right.
5 There should be no calculations in the table, only calculated values.
6 Each column should have an informative heading with the correct SI units placed in brackets.
7 Units should appear only in the column headings, not in the body of the table.
8 All raw data should be recorded to the same number of decimal places and significant figures appropriate to the least accurate piece of equipment used to measure it.
9 All processed data may be recorded to one decimal place more than the raw data.

Analysis

REVISED

As part of the written examination, you may need to:

1 process, analyse and interpret qualitative and quantitative experimental results
2 use appropriate mathematical skills for the analysis of quantitative data
3 use significant figures appropriately
4 plot and interpret suitable graphs from experimental results, including (i) selection and labelling of axes with appropriate scales, quantities and units, and (ii) measurement of gradients and intercepts

Processing

This involves calculating means, rates of reaction or some other aspect of the data. It also involves being able to draw a suitable graph. The type of graph used should be appropriate to the data collected. For all graphs:

- the graph should be of an appropriate size to make good use of the paper and include an appropriate scale
- there should be an informative title
- the axes must be the correct way around (independent variable on the horizontal axis), labelled and with units in brackets

Bar charts

1 Used when the independent variable is discontinuous or non-numerical, e.g. colours, species.
2 Data presented as lines or blocks of equal width that do not touch.
3 Lines or blocks can be arranged in any order but could be ascending or descending.

Histograms

1 Used when the independent variable is continuous and numerical.
2 Blocks should be drawn touching.
3 The y-axis represents the number in each class or the frequency.

Pie charts

1 Used when displaying data that are proportions or percentages.
2 Should not contain more than six or seven sectors to avoid confusion
3 Segments should be labelled or a key provided.

Line graphs

1 Use a true origin (0,0) or, if the origin is not included, the axis should be broken.
2 Points should be plotted with saltire crosses (×) or encircled dots.
3 Draw a smooth curve as a line of best fit.
4 If more than one curve is drawn on the same axes, each curve must be labelled to show what it represents.

Analysing

Use one of the statistical tests available. You should know when to use each of these tests:

1 the chi-squared test
2 a correlation coefficient
3 the Student's t-test

Use of significant figures

Any column in a table should have all the figures expressed to the same number of decimal points. The number of figures after the decimal point indicates precision in your measurements — this must be appropriate to the techniques or apparatus used. A mean or other processed data can be given to one extra figure after the decimal point.

Interpreting

1 Describe the trend shown in the data.
2 Make deductions and conclusions from your graphical data.
3 Justify these deductions by referring to the data.

Evaluation

REVISED

You need to evaluate your results and draw conclusions. This includes:
1 identifying anomalies in experimental measurements
2 identifying limitations in experimental procedures
3 understanding the meaning of precision and accuracy in the measurement of results
4 appreciating the margin of error, percentage errors and uncertainties in apparatus
5 suggesting improvements to experimental design, procedures and apparatus

Identifying anomalies

An anomaly is a result that does not fit the pattern shown in your other measurements. It may be caused by:
1 a procedural error, such as diluting a solution incorrectly
2 a timing error, such as leaving the reaction to continue for longer than the other readings
3 misreading the scale on the apparatus
4 living material that does not behave consistently
5 not controlling some other factor that affects the results (this is called a limitation)

Including an anomaly in the calculated mean makes the mean less reliable.

Limitations

Limitations are factors that may affect the results but have not been controlled or accounted for in some way. Examples include:
1 small data sets
2 variables that cannot be controlled
3 the degree of precision of the instruments used
4 the accuracy of the instruments used
5 insufficient repeats
6 insufficient time for acclimatisation or equilibration before taking measurements

Precision and accuracy

Precision is the ability to be exact and it depends on the ability of the equipment used and the units of measurement. Accuracy is how close your measurement comes to the true value.

Reliability

Reliability is how much confidence you have in the results. This depends on the errors and uncertainty. Errors can be caused by:

1 unsuitable practical procedures
2 poor judgement by the experimenter

Uncertainty is a measure of the precision and accuracy of the apparatus used. The level of uncertainty can be taken as half the smallest unit used for measurement. If you are measuring length using a ruler marked in millimetres, the level of uncertainty is 0.5 mm. If you are using a digital scale, the level of uncertainty is one full unit.

Where a reading relies on the judgement of the experimenter, such as when a colour has changed, the level of uncertainty (and therefore the level of reliability) is much larger and hard to quantify.

Reliability can be assessed by making repeat measurements. This enables the experimenter to:

1 identify anomalies
2 calculate a mean
3 calculate standard deviation
4 calculate percentage error or percentage uncertainty:

$$\text{Percentage uncertainty} = \frac{\text{uncertainty}}{\text{actual measurement made}} \times 100$$

Using range bars and error bars

Your data can be used to assess the reliability of your results. Your measurements of the dependent variable will be spread around the mean.

The range is the difference between the highest and the lowest figures. The greater the range, the less reliable your results. For example, if you take three readings of the dependent variable (34, 36 and 32), the mean is:

$$\frac{34 + 36 + 32}{3} = 34$$

The range is 32 to 36.

Range bars are vertical bars drawn on your graph to show the range of your results for a particular value of the independent variable.

The standard deviation of your results is a measure of the spread of the results. It is given by the formula:

$$s = \sqrt{\frac{\sum(x - \bar{x})^2}{n - 1}}$$

where s = standard deviation, $\bar{x}$ = the mean value and n = the number of data points. You can use the formula to calculate the standard deviation of your results. A larger standard deviation indicates a greater spread of results and less reliable data.

Error bars are drawn by plotting the values of the mean plus or minus the standard deviation (Figure 1.1).

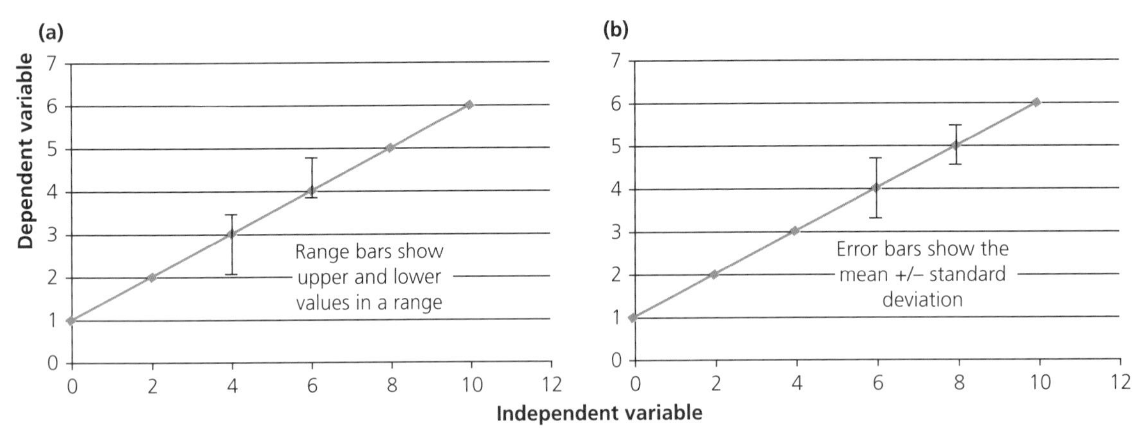

Figure 1.1 Graphs showing (a) range bars and (b) error bars

Summary

Practical skills will be tested in the written examination.

Planning may involve:

- solving a problem in a practical context
- selection of suitable apparatus, equipment or techniques
- describing a method or technique
- application of scientific knowledge
- identification of all variables
- evaluation of the method

Implementing may involve:

- describing how to use a wide range of practical apparatus and techniques
- suggesting precautions that should be taken to ensure results are valid
- assessing the hazards and potential risks involved
- deciding on appropriate units for measurements
- presentation of observations and data in an appropriate format (table)

Analysis may involve:

- processing qualitative and quantitative experimental results
- interpreting qualitative and quantitative experimental results
- using appropriate mathematical skills
- using significant figures appropriately
- plotting and interpreting suitable graphs from experimental results, including (i) selection and labelling of axes with appropriate scales, quantities and units, and (ii) measurement of gradients and intercepts
- drawing conclusions from data

Evaluation may involve:

- identifying anomalies in experimental measurements
- identifying limitations in experimental procedures
- understanding the meaning of precision and accuracy in the measurement of results
- appreciating the margin of error, percentage errors and uncertainties in apparatus
- calculating the standard deviation of a data set
- assessing the reliability of data using range bars and error bars
- calculating percentage error
- suggesting improvements to experimental design, procedures and apparatus
- assessing the validity of data

Exam practice

1 What is the most appropriate unit to measure the length of a mitochondrion? [1]

A nm **B** μm **C** mm **D** m

2 Results can be displayed in a number of different graphical forms. Which row in the following table correctly identifies the best way to display each set of data? [1]

Row	The effect of temperature on enzyme activity	A comparison of biodiversity in two fields	A belt transect across a ditch	A survey of the length of leaves on an oak tree
A	Histogram	Kite graph	Histogram	Bar chart
B	Line graph	Histogram	Kite graph	Pie chart
C	Line graph	Pie chart	Kite graph	Histogram
D	Line graph	Pie chart	Kite graph	Bar chart

3 A student investigated the effect of light intensity on the rate of transpiration from a leafy shoot using a potometer. Which row in the following table correctly identifies the variables? [1]

Row	Variables			
	Independent	Dependent	Control	Control
A	Light intensity	Rate of transpiration	Temperature	Air movement
B	Rate of transpiration	Light intensity	Temperature	Air movement
C	Light intensity	Air movement	Temperature	Rate of transpiration
D	Temperature	Rate of transpiration	Light intensity	Air movement

4 You are asked to make drawings from a slide of a plant leaf transverse section.

(a) (i) What type of drawing should be made using a 4× objective lens? [1]

(ii) Leaves contain palisade cells. List three other features you may include in your drawing. [3]

(b) Using a 10× eyepiece and a 40× objective lens you could draw individual plant cells.

(i) What is the total magnification used? [1]

(ii) List three features that you may include in your drawing of a palisade cell. [3]

5 A student investigated the effect of temperature on the rate of decomposition of hydrogen peroxide by yeast. He collected the gas released in a measuring cylinder. The following table shows the results of his investigation.

Gas produced (cm^3)			Temperature (°C)
Trial 1	Trial 2	Mean	
39	62	50.5	20
53	65	59	25
78	72	75	30
102	97	99.5	35

(a) Outline how the student may have carried out this investigation. [5]

(b) State two precautions that he should have taken to ensure the results were reliable. [2]

(c) Identify three ways in which the table of results could be improved. [3]

(d) Identify one anomalous result in the table and justify your choice. [3]

(e) Suggest two limitations to this investigation. [2]

(f) Suggest two ways in which the results collected could have been improved. [2]

Answers and quick quiz 1 online

ONLINE

2 Cell structure

Microscopes and images

There is a range of microscopes for creating images of cells and ultrastructure. Each microscope has different limits of magnification and resolution.

Magnification and resolution

REVISED

Magnification is the ratio of image size to object size (size of image/size of object). **Resolution** is the ability to distinguish between two objects that are close together — the ability to provide detail in the image.

$$\text{Magnification } (M) = \frac{\text{image size } (I)}{\text{actual size } (A)}$$

Exam tip

Every biology examination paper includes a calculation. This may be to do with magnification or image size. You need to be able to calculate the magnification, but you also need to be able to manipulate the formulae.

Magnification is the ratio of the image size to the object size.

Resolution is the ability to distinguish two separate points that are close together.

Revision activity

Make up an acronym to help you remember the magnification formula.

Now test yourself

1 Using the formula $M = \frac{I}{A}$, rearrange the letters to give the formulae for I and for A.

Answer on p. 222

TESTED

Four types of microscope are commonly used, as shown in Table 2.1. Their advantages and disadvantages are given in Table 2.2.

Table 2.1 Types of microscope

Type of microscope	Magnification	Resolution	Use
Light	1000–2000×	50–200 nm	Viewing tissues and cells
Scanning electron	50 000–500 000×	0.4–20 nm	Viewing the surface of cells and organelles Providing depth/three-dimensional images
Transmission electron	300 000–1 000 000×	0.05–1.0 nm	Detailing organelles (ultrastructure)
Laser scanning confocal	1000–2000×	50–200 nm	Three-dimensional images with good depth selection

Now test yourself

TESTED

2 Explain why a light microscope will not usually magnify images to greater than 1500×.
3 How can you tell the difference between an image created by a scanning electron microscope and one created by a transmission electron microscope?
4 Explain the relationship between millimetres (mm), micrometers (μm) and nanometers (nm).

Answers on p. 222

Table 2.2 Advantages and disadvantages of different microscopes

Type of microscope	Advantages	Disadvantages
Light	Cheap and easy to use Allows us to see living things	The resolution is limited
Scanning electron and transmission electron	The resolution is better than a light microscope, which means it is worth magnifying the image more as the image will show more detail Scanning electron microscopes also: ● give 3D images with depth of field ● are good for viewing surfaces	Electron microscopes are large and very expensive They require trained operatives The sample must be dried out and is therefore dead. This may affect the shape of the features seen (called an artefact) The image is in black and white, but colours may be added later by computer graphics. These are called false colour electron micrographs
Laser scanning confocal	Can also see living things and have the advantage that they can focus at a specific depth so the image is not confused by other components that are not in focus	Relies on a computer to piece together all the information from the dots of light created by the lasers. This means that the image is an interpretation rather than a real-life image

Staining

REVISED

Most cell components are colourless and hard to see. **Staining** is the application of coloured stains to the tissue or cells. Staining:

1. makes objects visible in light microscopes
2. increases contrast so that the object can be seen more clearly
3. is often specific to certain tissues or organelles. For example, acetic orcein stains chromosomes dark red, eosin stains cytoplasm, Sudan red stains lipids, and iodine in potassium iodide solution stains the cellulose in plant cell walls yellow and starch granules blue/black

In an electron microscope, the stains are actually heavy metals or similar atoms that reflect or absorb the electrons.

Revision activity

Draw a mind map to show the reasons for staining cells and tissues. Include the names of any stains you may have used such as methylene blue and iodine.

Cells

Cells are the basic unit of living organisms. All eukaryotic cells share a similar basic structure containing membrane-bound organelles. Each organelle, whether membrane-bound or not, has its own function within the cell.

Cell structure under the electron microscope

REVISED

You should be able to prepare cells and tissues to view under an electron microscope. This involves creating a smear or cutting a very thin section. These cells can then be stained appropriately and covered by a cover slip. Viewing these cells, interpreting what you can see and drawing what you see are important skills, as shown in the examples in Figure 2.1. Remember that details are important.

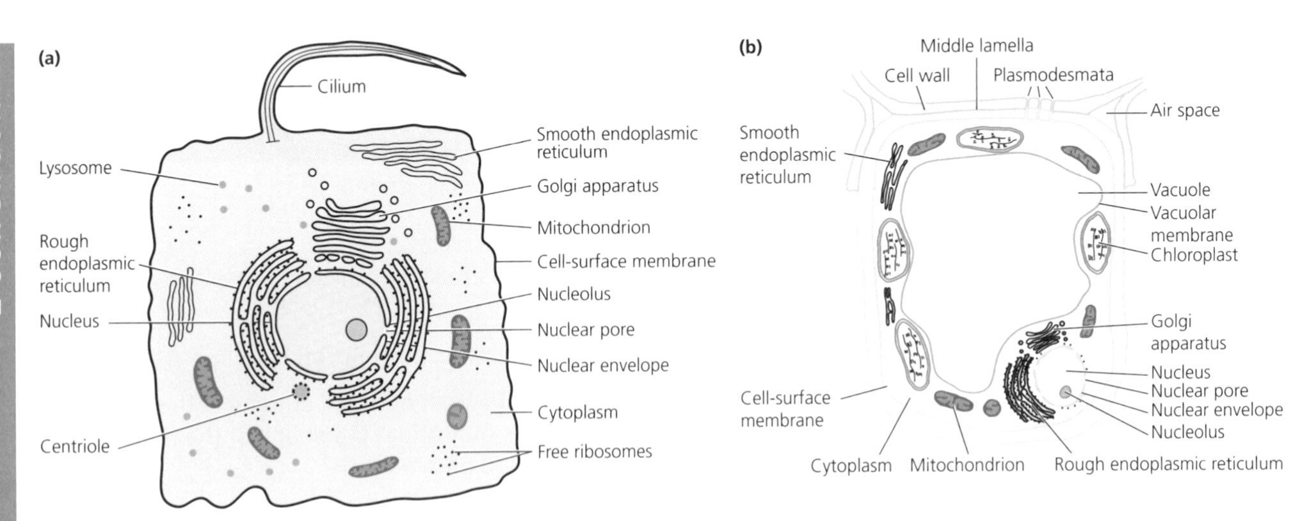

Figure 2.1 Diagrams of (a) an animal cell and (b) a plant cell as seen using an electron microscope

Cell ultrastructure

REVISED

The **ultrastructure** of a cell is the detail you can see using an electron microscope (Table 2.3).

Table 2.3 Organelles and their functions

Organelle	Function	Diagram
Cell wall	Cellulose cell wall surrounds plant cells	
Centrioles	Involved in the organisation of the microtubules that make up the cytoskeleton Form the spindle fibres used to move chromosomes in nuclear division	
Chloroplasts	Site of photosynthesis	
Cilia	Small hair-like extension of cell-surface membrane containing microtubules Large numbers work in synchronised fashion Able to move whole organism or to move fluid (mucus) across a surface	
Cytoskeleton	A network of microtubules and microfilaments Provides mechanical strength, support and structure for the cell Maintains the cell shape and is used in some cells to change the shape of the cell Enables movement of organelles inside the cell Enables movement of the whole cell	

Organelle	Function	Diagram
Flagella (undulipodium in eukaryotes)	Large extension of cell-surface membrane containing microtubules (in eukaryotes) Able to beat to enable locomotion or move fluids	
Golgi apparatus	Modifies proteins made in the ribosomes Often adds a carbohydrate group Repackages proteins into vesicles for secretion	
Lysosomes	Small vacuoles containing hydrolytic or lytic enzymes	
Mitochondria (singular: mitochondrion)	Site of aerobic respiration	
Nucleus, nucleolus and nuclear envelope	Contains the genetic material (chromosomes) Controls the cell activities The nuclear envelope separates the genetic material from the cytoplasm The nuclear pores allow molecules of mRNA to pass from the nucleus to the ribosomes in the cytoplasm The nucleolus assembles the ribosomes	
Ribosomes	Site of protein synthesis	
Rough endoplasmic reticulum (RER)	Holds many of the ribosomes Provides a large surface area for protein synthesis	
Smooth endoplasmic reticulum (SER)	Associated with the synthesis, storage and transport of lipids and carbohydrates	

Now test yourself

TESTED

5 Explain why organelles such as mitochondria do not always look the same size and shape.

Answer on p. 222

Revision activity

From memory, make a list of all the membrane-bound organelles and note one function next to each. Make a separate list of the organelles that are not membrane-bound.

How the organelles work together

REVISED

The **organelles** in a cell work together to achieve the overall function of that cell. Many of the organelles are involved in the production and secretion of **proteins**. The sequence of events always follows the same course:

1. mRNA leaves the nucleus via the nuclear pores.
2. It is used by the ribosomes on the rough endoplasmic reticulum to construct a protein.
3. The protein travels in a vesicle to the Golgi apparatus.
4. The vesicle is moved by the **cytoskeleton**, possibly using tiny protein motors that 'walk' along the microtubules using them as a track.
5. The Golgi apparatus modifies the protein (often adding a carbohydrate group) and repackages it into a vesicle.
6. This vesicle is moved to the cell-surface (plasma) membrane.
7. The vesicle fuses with the membrane to release the protein from the cell.

Revision activity

Draw a flow diagram of the sequence of events leading to the secretion of a protein.

Exam tip

Always remember to say that the plasma membrane is involved in secretion and that the vesicle fuses to this membrane.

Prokaryotic and eukaryotic cells

REVISED

There are two types of cell: **prokaryote** (Figure 2.2) and **eukaryote** (Figure 2.1a). The features of each type are given in Table 2.4.

Table 2.4 Comparing prokaryotic and eukaryotic cells

Feature	Prokaryote	Eukaryote
Size	Smaller — typically less than 10 μm long and 1–2 μm wide	Larger — typically larger than 10 μm in diameter
Nucleus	No	Yes
Membrane-bound organelles	No	Yes
Ribosomes	Yes, 18 nm in size	Yes, 22 nm in size
Chromosomes	A single loop of DNA, no histones	DNA associated with proteins (histones)
Flagellum	Some cells have a flagellum. It has a very different structure	Some cells have a flagellum with 9 + 2 structure of microtubules

Revision activities

- Make a list of the organelles found in a plant cell.
- Make a list of the organelles found in an animal cell.
- List the features of a bacterial cell that are also seen in plant cells.

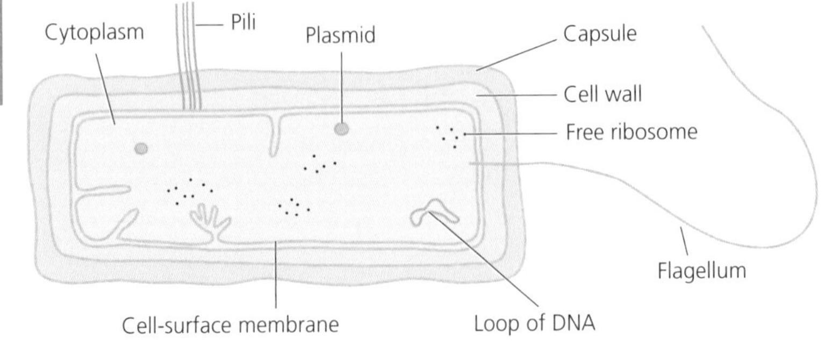

Figure 2.2 The generalised structure of a prokaryotic cell

Revision activity

Write a list of all the key terms used in this chapter, then add the meaning of each key term.

Exam practice

1 Which row in the following table correctly identifies the sequence of cell components used during the secretion of a protein? [1]

Row	Sequence			
A	Golgi apparatus	Vesicles	Ribosomes	Plasma membrane
B	Vesicles	Ribosomes	Golgi apparatus	Plasma membrane
C	Plasma membrane	Vesicles	Golgi apparatus	Ribosomes
D	Ribosomes	Vesicles	Golgi apparatus	Plasma membrane
E	Golgi apparatus	Vesicles	Ribosomes	Plasma membrane

2 Which row in the following table correctly describes the components found in prokaryotic and eukaryotic cells? [1]

Row	Prokaryotic cells	Eukaryotic cells
A	Mitochondria, chloroplasts, ribosomes	Mitochondria, chloroplasts, ribosomes
B	Ribosomes, plasmids, flagellum	Nucleus, ribosomes, mitochondria
C	Mitochondria, chloroplasts, ribosomes	Nucleus, ribosomes, mitochondria
D	Nucleus, plasmids, ribosomes	Flagellum, chloroplasts, nucleus

3 The following statements are about the organelles in cells.

A Mitochondria are where respiration occurs.
B Chloroplasts contain chlorophyll.
C Cilia are found only in the airways.
D The cytoskeleton helps cells to change shape.
E Ribosomes are larger than mitochondria.

Which of the following options identifies the correct statements? [1]

(a) All five statements are correct.
(b) Only statements A, B and C are correct.
(c) Only statements A, B and D are correct.
(d) All five statements are incorrect.

4 Complete the following paragraph. [5]

The four main types of microscope are ...., scanning electron, transmission electron and laser scanning confocal. The advantage of an electron microscope is that it has much greater .... than other types of microscope. This allows the user to see far more detail. A transmission electron microscope is used to see the detail of .... whereas a scanning electron microscope gives detail of the cell .... and a three-dimensional image. The advantage of a laser scanning confocal microscope is that it can produce images at a specific .... within the tissue or cell.

5 A student prepared a slide of onion epidermis using the stain methylene blue.

(a) Explain the advantages of staining a tissue such as onion epidermis. [2]
(b) The student drew one cell. Calculate the magnification of the image. [2]

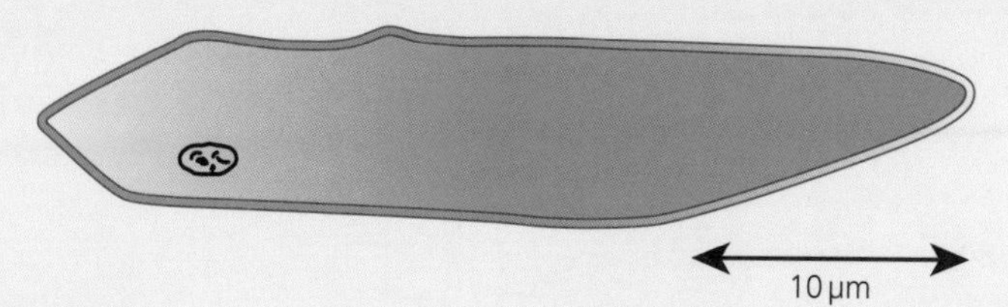

(c) What feature shown in the student's diagram indicates that the cell is eukaryotic? [1]

6 (a) Complete the following table to show organelles and their functions. [4]

Organelle	Function
Mitochondrion	
Chloroplast	
	Modify proteins
	Manufacture proteins

(b) What is the role of the nuclear pores? [2]
(c) Describe how the structure of a mitochondrion is adapted to its function. [2]
(d) Suggest why lytic enzymes are held inside lysosomes. [2]
(e) Describe the roles of the cytoskeleton. [5]

Answers and quick quiz 2 online

Summary

By the end of this chapter you should be able to:

- Know the differences between light microscopes, electron microscopes and laser scanning confocal microscopes and describe the advantages and disadvantages of each.
- Appreciate the difference between resolution and magnification.
- Be able to calculate magnifications or true sizes given sufficient information.
- Understand that staining increases contrast and makes parts of the specimen stand out so that they can be seen more clearly.
- Differential staining allows different molecules, organelles or tissues to be stained different colours.
- Be able to recognise each type of organelle from a photomicrograph and from a drawing.
- Know what function each type of organelle has in the cell.
- Remember that most organelles consist of membranes and that the nucleus, the mitochondria and the chloroplasts have two membranes.
- Understand that membranes inside cells are distinct from the plasma (cell-surface) membrane that surrounds the cytoplasm.
- Be able to describe the sequence of events leading to the secretion of a protein from the cell.
- Know the differences between prokaryotic and eukaryotic cells.

3 Biological molecules

Water

Hydrogen bonds

REVISED

Water is a simple molecule that can form **hydrogen bonds** between its molecules. This gives water some very important properties.

Hydrogen bonds are weak forces of attraction. They can form between molecules or between parts of a larger molecule. In water there is attraction between the oxygen of one molecule and the hydrogen of another molecule.

This attraction occurs because each water molecule is polar, which means there is an uneven distribution of charge. Oxygen atoms attract electrons more strongly than hydrogen atoms. Therefore, the electrons in a water molecule are pulled towards the oxygen atom, which gives the oxygen end of the molecule a more negative charge. This is shown as delta negative (δ^-). The hydrogen end of the molecule is left with a delta positive (δ^+) charge. It is these opposite delta charges that attract the water molecules together, producing hydrogen bonds (Figure 3.1).

Hydrogen bonds are weak forces of attraction between molecules or parts of a molecule that are polar.

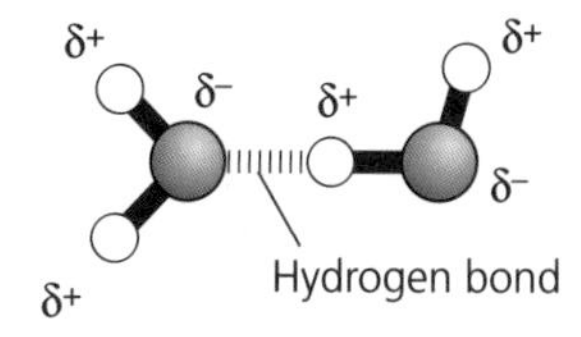

Figure 3.1 Water molecules are polar and are held together by hydrogen bonds

Hydrogen bonds and the properties of water

REVISED

Hydrogen bonds cause cohesion. Between 0°C and 100°C, they hold water molecules together loosely — they are held together, but they can move past one another and the water remains a liquid. In order to evaporate, the hydrogen bonds must be broken, allowing the molecules to separate and form a gas (water vapour). This takes a lot of energy, so water remains a liquid up to 100°C.

At lower temperatures the molecules have less kinetic (movement) energy and move about less. With less movement, more hydrogen bonds can form and at 0°C enough hydrogen bonds have formed to hold the water molecules in a stationary position, forming ice. The water molecules are now held in a rigid lattice, which holds the molecules further apart. Therefore, ice floats as it is less dense than water.

Exam tip

In any response, ensure you use the hydrogen bonds between the molecules of water to explain its properties.

Typical mistake

Many candidates fail to make the link between the action of the hydrogen bonds and the properties of water.

Water properties and living things

REVISED

The following properties of water are important for the survival of many living things:

- Thermal stability — water has a high specific heat capacity, which means that a lot of energy is needed to warm it up. Therefore, a body of water maintains a fairly constant temperature, which is essential for life to survive.
- Freezing — ice is less dense than water so it floats, which insulates the water and prevents it freezing completely. Living things can survive below the ice.

Typical mistake

Many candidates fail to indicate how the properties of water are important to the survival of living things.

- Evaporation — a lot of energy is needed to cause evaporation, which is used to cool the surface of living things. This energy is known as latent heat. Water has a high specific latent heat capacity.
- At most temperatures water is a liquid — it can flow and transport materials in living things.
- **Cohesion** — the attraction of water molecules produces surface tension, which creates a habitat on the surface. It also enables continuous columns of water to be pulled up the xylem.
- Solvent — as the molecules are polar, water can dissolve a wide range of substances.
- As a reactant — water molecules are used in a wide range of reactions such as hydrolysis and photosynthesis.
- Incompressibility — water cannot be compressed into a smaller volume. This means it can be pressurised and pumped in transport systems or used for support in hydrostatic skeletons.

Cohesion is the weak attraction between two similar molecules.

Revision activity

Draw a mind map with water in the centre to show how its properties are essential for living things.

Now test yourself

TESTED

1 Explain what is meant by a δ^+ charge and how it is created.

2 (a) Explain the difference between specific heat capacity and latent heat capacity.

(b) What role does each play in the survival of living things?

Answers on p. 222

Monomers and polymers

The concept of monomers and polymers

REVISED

Many molecules are **polymers**. These are long chains that comprise a number of similar smaller molecules known as **monomers**. There are three biologically important groups of polymers found in living organisms: nucleic acids, polysaccharides and proteins.

In nucleic acids, the monomers are nucleotides. They are made up of a:

1. five-carbon or pentose sugar. This is either ribose (in the case of RNA) or deoxyribose (in DNA)
2. phosphate group
3. nucleotide base. There are five different bases. These are adenine, cytosine, guanine, thymine and uracil, which are usually abbreviated to A, C, G, T and U. The nucleotides that make up DNA contain one of the bases adenine, cytosine, guanine or thymine. In RNA, the thymine is replaced by uracil

Complex carbohydrates such as starch and cellulose are polysaccharides — a group of polymers made up of monosaccharides.

In proteins, the monomers are amino acids. There are 20 different amino acids used to build proteins. The different sequences that these amino acids can be used in provide huge diversity in protein structure.

Polymers are long chains of repeated units. The individual units are called monomers.

Monomers are one of the small similar molecules that join together to form a polymer. There are three important biological monomers: amino acids, monosaccharides and nucleotides.

Condensation and hydrolysis

REVISED

The monomers in both complex carbohydrates and proteins are joined by covalent bonds created by a **condensation** reaction (Figure 3.2). This is a reaction that releases a water molecule. The bond can be broken again by adding a water molecule. This is called **hydrolysis**.

Condensation is a reaction that releases a molecule of water.

Glycine

Alanine

Hydrolysis H_2O ⇌ Condensation H_2O

Dipeptide

Peptide bond

Figure 3.2 Condensation forms a peptide bond between two amino acids. Hydrolysis can break the bond

> **Hydrolysis** is splitting a large molecule into two smaller molecules by the addition of water.

The chemical elements that make up biological molecules

Carbohydrates

REVISED

Carbohydrates consist of carbon, oxygen and hydrogen. The general formula for a carbohydrate is $(CH_2O)_x$. Carbohydrates can be divided into three main groups: monosaccharides, disaccharides and polysaccharides.

Monosaccharides

These are the single sugar units that are used as monomers to build other carbohydrates. They are soluble and sweet reducing sugars.

Pentose sugars

Pentose monosaccharides contain five carbons, e.g. **ribose** and deoxyribose (Figure 3.3).

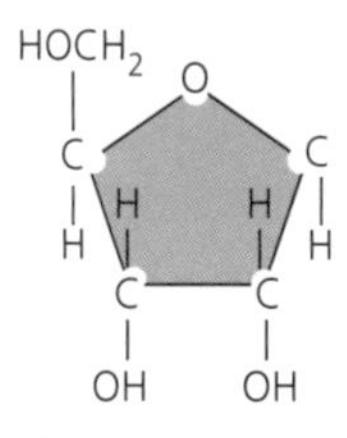

Figure 3.3 Ribose, a pentose sugar

Hexose sugars

Hexose monosaccharides contain six carbons, e.g. **glucose**. There are two forms of glucose: alpha (α) glucose and beta (β) glucose. The structures of α-glucose and β-glucose are shown in Figure 3.4.

The difference between α-glucose and β-glucose is simply the position of the –H and –OH groups on the first carbon atom. This difference is not great, but it has a large effect on the way they bond together and the polysaccharides produced.

(a)

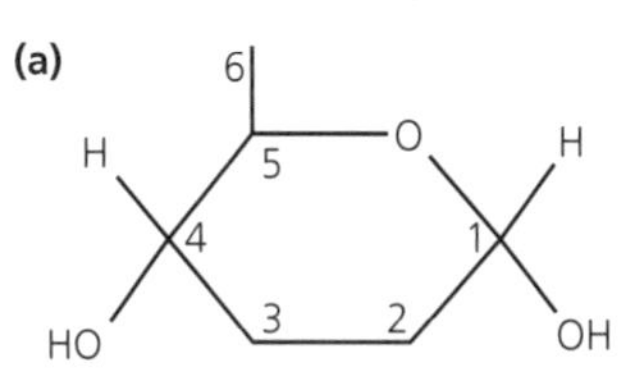

α-glucose (note that the –H is above the –OH on carbon atom 1)

(b)

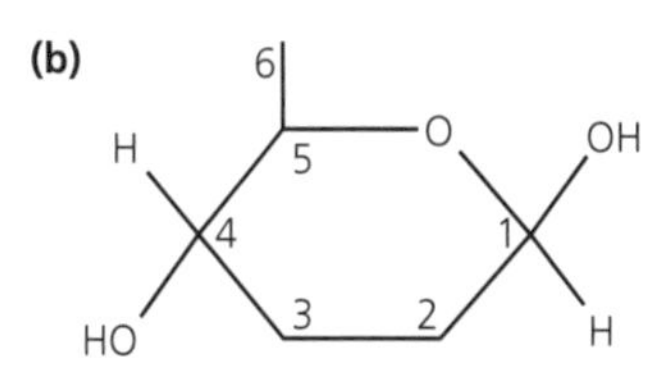

β-glucose (note that the –OH is above the –H on carbon atom 1)

Figure 3.4 (a) α-glucose and (b) β-glucose. Note that the carbon atoms are numbered from the oxygen clockwise

Disaccharides

Disaccharides are two monosaccharides bonded together. They are bonded by a covalent bond called a **glycosidic bond**. Disaccharides are soluble and sweet. Most are reducing sugars, but sucrose is not a reducing sugar.

The **glycosidic bond** is the bond between two monosaccharides in a polysaccharide.

Forming disaccharides

Disaccharides are formed by a condensation reaction between two monosaccharides. The most common bond is between the carbon 1 of one monosaccharide and the carbon 4 of another. This forms a 1,4 glycosidic bond and releases water.

Glucose and fructose combine to form sucrose, whereas glucose and glucose combine to form maltose (Figure 3.5), and glucose and galactose combine to form lactose. Disaccharides can be converted back to monosccharides by hydrolysis.

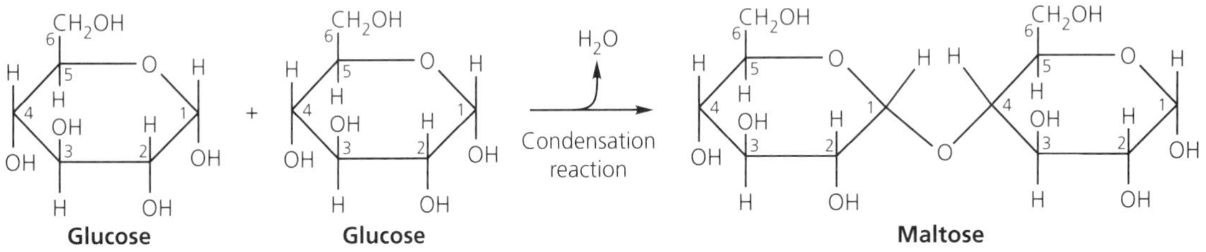

Glucose + Glucose —H_2O Condensation reaction→ Maltose

Figure 3.5 Two α-glucose molecules combine to form maltose

Polysaccharides

Polysaccharides are large insoluble molecules. **Starch** is actually a combination of two molecules, amylose and amylopectin.

Amylose

Amylose consists of a long unbranched chain of α-glucose subunits. The subunits are joined by 1,4 glycosidic bonds. The chain of subunits coils up. The hydroxyl group on carbon 2 of each subunit is hidden inside the coil. This makes the molecule less soluble.

Amylose is used for the storage of glucose subunits and energy in plant cells. The molecule is compact — it takes little space in the cell. It is insoluble, which means the molecules do not affect the water potential of the cells.

Glucose subunits can be removed easily from each end of the molecule. They can be used as building blocks to build other substances or as a substrate in respiration to release stored energy.

Amylopectin and glycogen

Amylopectin and **glycogen** are similar to amylose in that they are both long chains of α-glucose subunits bonded by 1,4 glycosidic bonds (Figure 3.6). Some of the glucose subunits also have 1,6 glycosidic bonds as well as the 1,4 glycosidic bonds. This means that the molecule is branched.

Amylopectin occurs in plants and has few branches. Glycogen is used for storage of glucose subunits and energy in animal cells. Glycogen has more 1,6 glycosidic bonds, making it more branched. This means that the molecule has many ends from which glucose can be released quickly. It has the same advantages as amylose: it is insoluble and compact.

Exam tip

Complex carbohydrates are not that complex — they are simply long chains of one or two subunits. Each chain is called a polymer.

Exam tip

You need to know about amylose and amylopectin, but you should be able to apply that knowledge to other complex carbohydrates.

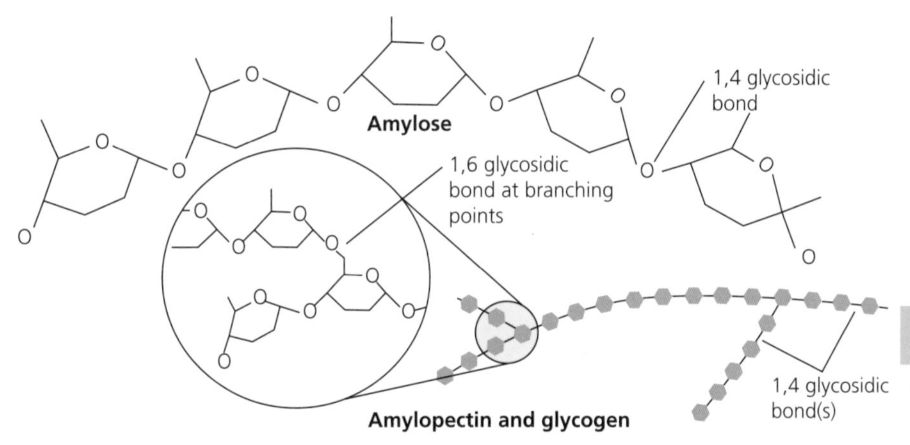

Figure 3.6 Amylose is a long, unbranched chain. Amylopectin and glycogen have branches formed by 1,6 glycosidic bonds

Cellulose

Cellulose consists of a long unbranched chain of β-glucose subunits. The subunits are joined by a 1,4 glycosidic bond. The chain of β-glucose subunits form a straight chain (Figure 3.7).

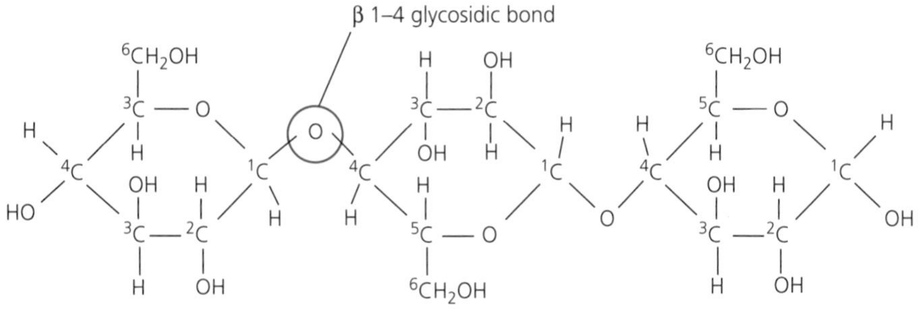

β 1–4 glycosidic bond

Figure 3.7

The hydroxyl groups on carbon 2 of each subunit are exposed, allowing hydrogen bonds to form between adjacent cellulose molecules. Some 60–70 molecules bind together to form a cellulose microfibril and many microfibrils join together to form macrofibrils.

Cellulose is strong and completely insoluble. It is used in plant cell walls and provides enough strength to support the whole plant.

Now test yourself

3 List the reasons why amylose is a good storage product.

Answer on p. 222

TESTED

Typical mistake

Many candidates are mistaken in thinking that cellulose is a protein, not a carbohydrate.

Exam tip

The structure of complex carbohydrates lends itself to a question in which the examiner asks you to relate the structure of the molecule to the function of that molecule. This is perhaps most easily done as a table.

Now test yourself

TESTED

4 Explain why cellulose is insoluble.

Answer on p. 222

Revision activity

Draw a table to compare the structures and properties of amylose, glycogen and cellulose.

Lipids

REVISED

Lipids are not polymers like proteins and complex carbohydrates. They are a large group of compounds that includes triglycerides, phospholipids and steroids. Lipids are insoluble in water.

Triglycerides

A **triglyceride** is a **macromolecule** containing one glycerol molecule and three fatty acid chains (Figure 3.8). The fatty acids are attached to the glycerol by a condensation reaction. The bonds are called **ester bonds** and they can be broken by hydrolysis.

A **triglyceride** is a molecule that comprises one glycerol molecule and three fatty acid chains.

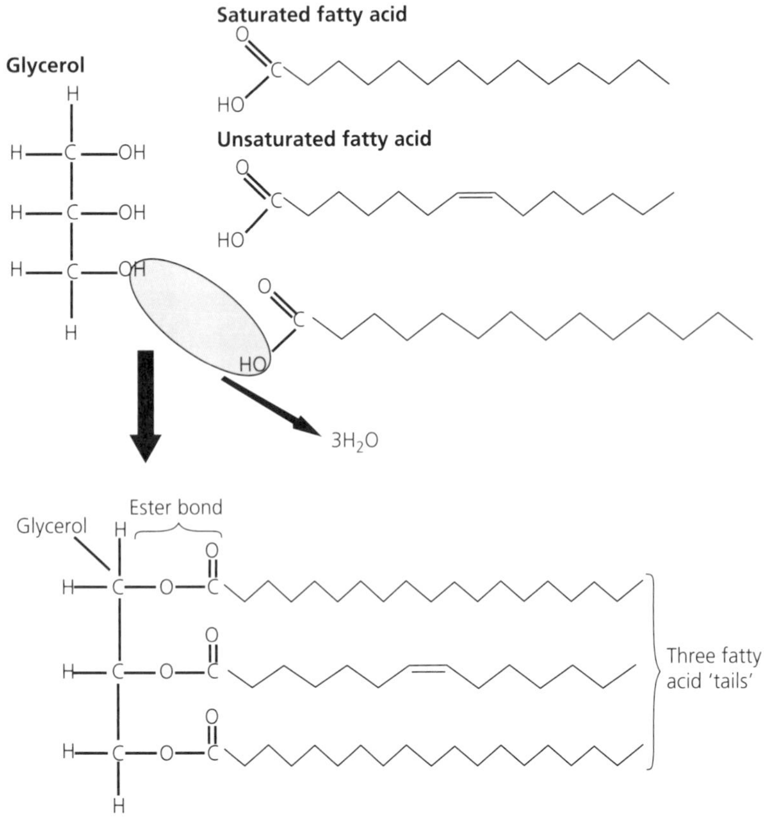

Saturated fatty acid

Glycerol

Unsaturated fatty acid

$3H_2O$

Ester bond

Glycerol

Three fatty acid 'tails'

Figure 3.8 Three fatty acids combine with one glycerol to produce a triglyceride. The elimination of one water molecule is shown. Three water molecules are released

Triglyceride molecules are rich in energy and used to store excess energy. When required, the molecules can be broken down in aerobic respiration to release this energy. Water is also released, which can be useful for animals that live in dry environments — hence camels store fat in their humps. The stores can be held under the skin and around major organs. It has the benefit of protecting the major organs from physical shock.

Triglycerides are also good insulators and are used to insulate animals that live in cold environments such as polar bears and aquatic mammals such as whales. They also provide buoyancy for these mammals.

Saturated and unsaturated fatty acids

Fatty acids are long chains of carbon atoms with hydrogen atoms bonded to them. If each carbon has two hydrogen atoms attached, there are no double or triple bonds in the fatty acid. This is called a saturated fatty acid. Saturated fatty acids are found in animal fats. They have a higher melting point and are more solid at room temperature, like butter.

If there are fewer hydrogen atoms, there will be double or even triple bonds between adjacent carbon atoms. This is called an unsaturated fatty acid. Unsaturated fatty acids are found in plant fats and oils. They have lower melting points and are more likely to be liquid at room temperature, like spreads and vegetable oil.

Now test yourself

5 Explain why lipids are good storage molecules.

Answer on p. 222

TESTED

Phospholipids

Phospholipids are similar to triglycerides, but one of the fatty acid chains is replaced by a phosphate group (Figure 3.9). The two remaining fatty acid 'tails' are insoluble in water and are called **hydrophobic**.

A **phospholipid** is a molecule containing one glycerol molecule, a phosphate group and two fatty acid chains.

Hydrophobic means water-hating, repelled by water.

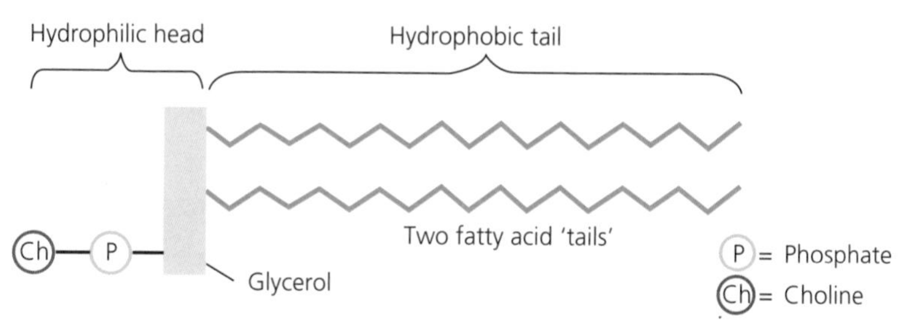

Figure 3.9 A phospholipid has two fatty acid chains and a phosphate group

The phosphate group is complex and includes choline, which is water soluble. This alters the characteristics of the molecule. This group makes the 'head' end of the phospholipid able to mix with water — it is **hydrophilic**. Phospholipids form bilayers with the hydrophobic 'tails' in the centre and the hydrophilic 'heads' pointing outwards to interact with the surrounding aqueous solution. This is the basis of all cell membranes.

Hydrophilic means water-loving, attracted to water.

Revision activity

Write an exam-style question worth about 10 marks about the roles of lipids in living things. Ensure that at least 5 marks test assessment objective 2 (use of your knowledge in an unfamiliar context). Write out a mark scheme.

Now test yourself

TESTED

6 Explain why phospholipids are used in membranes.

Answer on p. 222

Amino acids and proteins

REVISED

Amino acids

Proteins are made up of long chains of **amino acids**. There are 20 different amino acids used in proteins, but all have the same basic structure (Figure 3.10). The residual *R* group is the only part that differs between different amino acids.

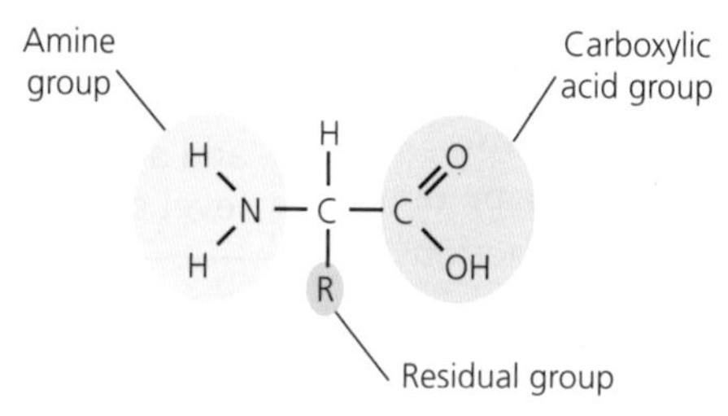

Figure 3.10 The generalised structure of an amino acid

Proteins

All proteins consist of long, unbranched chains of amino acids, which are held together by **peptide bonds**. These bonds are formed by **condensation** and occur between the amine group of one amino acid and the carboxylic acid group of another. A peptide bond is formed by condensation.

Two amino acids together make a **dipeptide** (Figure 3.11). Many amino acids in a chain form a **polypeptide**.

A **peptide bond** is the bond between two amino acids.

Condensation is a reaction that involves the release of water molecules.

A **dipeptide** is formed when two amino acids are joined together by a peptide bond.

A **polypeptide** is a chain of many amino acids joined together by peptide bonds.

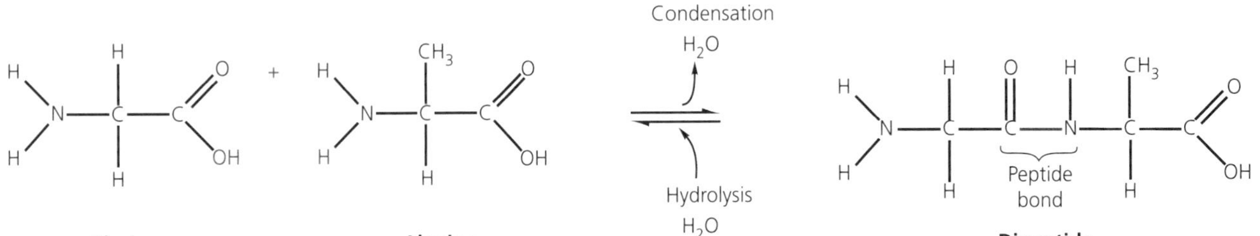

Glycine + Alanine ⇌ Dipeptide (Condensation H_2O / Hydrolysis H_2O; Peptide bond)

Figure 3.11 The condensation reaction to form a dipeptide

There are four levels of **protein structure** (Figure 3.12).

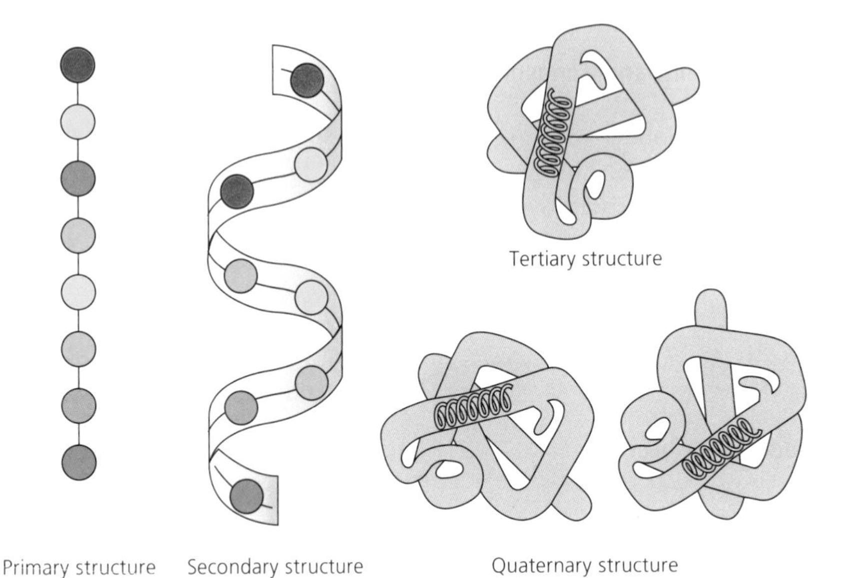

Figure 3.12 The four levels of structure in a protein

A **primary structure** is a chain of amino acids, held together by peptide bonds.

A **secondary structure** is formed when the chain of amino acids becomes folded and coiled. Two shapes are formed:

- alpha (α) helix — shaped like a coil spring
- beta (β) sheets — pleated like a folded sheet of paper

Hydrogen bonds hold the folds and coils in place.

A **tertiary structure** is formed when the coiled and pleated chains can be folded further to produce the final three-dimensional shape of the molecule. These final folds and coils are caused by the interactions between the *R* groups on the amino acids, which interact to form a range of bonds that hold the three-dimensional shape. These bonds include:

- hydrogen bonds between polar *R* groups
- ionic bonds between *R* groups with opposite charges
- covalent disulfide bonds between two sulfur-containing *R* groups

Additionally, some *R* groups are hydrophobic and twist away from water into the centre of the molecule. Others are hydrophilic and twist outwards so that they are on the outside of the molecule.

Many proteins are just one polypeptide chain that has been coiled and folded. However, some proteins consist of more than one polypeptide chain — e.g. haemoglobin has four polypeptide chains and collagen has three. These multi subunit proteins make up the **quaternary structure**.

Now test yourself

7 Explain why proteins are unbranched chains.

Answer on p. 222

TESTED

Typical mistake

Some candidates suggest that a protein shows just one of the levels of structure or that the labelling applies to the nutritional value of the protein. This is not the case — all proteins show a primary, a secondary and a tertiary level of structure, and some show a quaternary level of structure.

Exam tip

A typical question might ask you to relate the structure of a protein to its properties — this could be answered in the form of a table.

Globular proteins

Globular proteins are proteins that are highly folded to form a globular shape. These proteins are more water soluble. They are active in metabolism and their activity relies on their three-dimensional shape. Their shape and activity are sensitive to temperature changes: higher temperatures can cause distortion of their shape.

Haemoglobin

Haemoglobin is a **conjugated protein**. It is used to transport oxygen in the form of oxyhaemoglobin. Haemoglobin contains four polypeptide chains called subunits — two alpha (α) chains and two beta (β) chains.

Haemoglobin is a conjugated protein as each subunit has a non-protein prosthetic group attached, called a haem group, which contains a single iron ion (Fe^{2+}). One oxygen molecule can attach to each haem group, so a haemoglobin molecule can carry four oxygen molecules.

Enzymes

An example of an **enzyme** is amylase, which hydrolyses the bonds between glucose subunits in amylose. The molecule has regions that are coiled in an alpha (α) helix and other regions that are folded into beta (β) sheets. The compact globular shape contains an active site that has a specific shape that is complementary to the shape of the substrate (in this case, amylose). The active site holds at least one calcium ion that acts as a cofactor — it is essential for the correct action of the enzyme.

Hormones

An example of a hormone is **insulin**, which is used to stimulate removal of excess glucose from the blood. There are two polypeptides held together by disulfide bridges. One polypeptide is 21 amino acids and the second is 30 amino acids. The molecule has a specific three-dimensional shape that is complementary to the shape of a glycoprotein receptor on the surface of cells in the liver.

Fibrous proteins

Fibrous proteins tend to have a regular sequence of amino acids that is repeated many times. They are less soluble in water (usually totally insoluble). They tend to form fibres that have structural functions and examples include collagen, keratin and elastin.

Collagen

Collagen has three polypeptide chains wound around one another. It is not easily stretched. It provides strength in the walls of arteries to withstand the high blood pressure. It is found in tendons, which hold muscle to bone, and in bone, where it is hardened by calcium phosphate.

Keratin

Keratin has two polypeptide chains coiled together. It is strong and is used for protecting delicate parts of the body. Examples include finger nails, claws, hooves, horns, scales, hair and feathers. The cells in the outer layer of skin also contain keratin, which makes them impermeable to water.

Now test yourself

8 Explain the difference between a polypeptide and a protein.

Answer on p. 222

TESTED

A **conjugated protein** is a globular protein with a prosthetic group.

Revision activity

Make a model (or draw a diagram) of a protein using coiled wire or a slinky spring and folded paper. Label and annotate the model to show examples of the different types of bonding.

Elastin

Elastin is produced by linking many tropoelastin fibres together. Tropoelastin is coiled like a spring and can stretch and recoil. It is used wherever stretching and recoil is required, such as in the walls of the arteries and airways, alveoli, skin and the wall of the bladder.

Now test yourself

9 Explain why proteins that are metabolically active such as enzymes or hormones are globular proteins.
10 Explain why globular proteins are more affected by temperature change than fibrous proteins.

Answers on p. 222

The key inorganic ions involved in biological processes

Inorganic ions

REVISED

Inorganic ions are charged particles that have a number of important roles. These roles range from creating skeletal structures to nervous conduction and activating enzymes. Cations are positively charged and anions are negatively charged. Table 3.1 summarises these roles.

Table 3.1 The roles of inorganic ions

Ion	Role in biological systems
Ammonium (NH_4^+)	A component of amino acids, proteins and nucleic acids Involved in the: ● nitrogen cycle ● maintenance of pH
Calcium (Ca^{2+})	Increases the hardness of bones, teeth and the exoskeletons of crustaceans. It is also found in the middle lamella between plant cells A factor in blood clotting Involved in the control of muscle contraction and synaptic action Activates enzymes such as amylase and lipase
Chloride (Cl^-)	Involved in the: ● reabsorption of water in the kidney ● regulation of water potentials in cells and body fluids ● transport of carbon dioxide in the blood ● production of hydrochloric acid in the stomach
Hydrogen (H^+)	Involved in: ● oxidative phosphorylation in respiration ● photophosphorylation in photosynthesis ● the transport of carbon dioxide in the blood ● the regulation of blood pH ● reduction reactions in metabolism
Hydrogencarbonate (HCO_3^-)	Involved in the: ● regulation of blood pH ● transport of carbon dioxide in the blood

Ion	Role in biological systems
Hydroxide (OH^-)	Involved in the regulation of blood pH
Iron (Fe^{2+})	Increases the affinity of haemoglobin for oxygen
Sodium (Na^+)	Involved in: ● the regulation of water potentials in cells and body fluids ● the selective reabsorption of sugars and amino acids in the kidney ● the reabsorption of water in the kidney ● nervous transmission and muscle contraction
Magnesium (Mg^{2+})	Found at the centre of chlorophyll
Nitrate (NO_3^-)	A component of amino acids, proteins and nucleic acids Involved in the nitrogen cycle
Phosphate (PO_4^{3-})	A component of phospholipids, ATP and nucleic acids Increases the hardness of bones, teeth and the exoskeletons of crustaceans Improves root growth in plants
Potassium (K^+)	Improves growth of leaves and flowers in plants Involved in the: ● regulation of water potentials in cells and body fluids ● nervous transmission and muscle contraction

Testing for the presence of biological molecules

Chemical tests

REVISED

The presence of each biological molecule can be detected by a specific test. Table 3.2 summarises the procedures to be followed.

Table 3.2 Chemical tests to identify the presence of biological molecules

Molecule tested for	Test name	Details of test	Positive result
Protein	Biuret test	Dissolve in water. Add biuret A and biuret B	Colour change from blue to mauve or purple
Reducing sugar (e.g. glucose)	Benedict's test	Dissolve in water. Add Benedict's reagent. Heat at 80–90°C for 2 minutes	A precipitate forms. Colour change from blue to brick red
Non-reducing sugar (e.g. sucrose)	Benedict's test (this test converts non-reducing sugars to reducing sugars and then tests the reducing sugars)	Test the substance for reducing sugars to ensure that none are present, then dissolve in water. Add a few drops of dilute hydrochloric acid and boil for 2 minutes. Neutralise the solution by adding a few drops of dilute sodium hydroxide. Add Benedict's reagent and reheat for 2 minutes	A precipitate forms and there is a colour change from blue to brick red
Starch (amylose)	Iodine solution test	Dissolve in water. Add iodine solution	Colour change to deep blue/black
Fat (lipids)	Emulsion test	Dissolve in alcohol. Filter. Add water to the filtrate	When water is added to the clear filtrate, it will turn cloudy or milky

Making the Benedict's test quantitative

When testing for reducing sugars such as glucose, the colour change may not be complete: the colour may show a change from blue to green or yellow before going orange or red. If only a small amount of sugar is present, it will not react all the Benedict's reagent — leaving some of it blue. This is what causes an incomplete colour change.

The **Benedict's test** can be made quantitative (i.e. you can determine the concentration of the reducing sugars) by ensuring there is excess Benedict's reagent. Create a range of colours using known concentrations of reducing sugars to create a set of colour standards.

To make the measurement more precise, the standards and the sample should be placed in a centrifuge and spun for 2 minutes. This will deposit the coloured precipitate at the bottom of the tube, leaving a blue solution of unused Benedict's reagent. The more concentrated the reducing sugar, the less blue colour will be left. The intensity of the blue colour in the solution can be measured using a **colorimeter**. Plot a graph of absorbance against concentration using the standard solutions. Use your graph to determine the concentration of the unknown solution by reading from the absorbance measurement across to the concentration.

Typical mistake

This is straightforward recall, yet many candidates don't seem to learn the details. You must learn all these tests.

Now test yourself

TESTED

11 Explain why carrying out the Benedict's test on a solution of glucose at low concentration will leave the solution blue or green in colour, but if the glucose is at high concentration the solution will have no sign of blue colouration.

Answer on p. 222

Biosensors

A **biosensor** converts a chemical variable into an electrical signal. For example, a glucose biosensor measures the concentration of glucose. When the biosensor is dipped into a solution, the glucose diffuses towards immobilised enzymes. These catalyse a reaction that releases hydrogen peroxide. The hydrogen peroxide reacts with a platinum electrode to generate a current. The current generated is proportional to the glucose concentration.

Chromatography

Chromatography is a technique used to separate molecules in a mixture.

Paper chromatography involves placing a small sample of the mixture solution on to a strip of chromatography paper. The paper is then placed with one end in a shallow layer of solvent. The solvent rises up through the paper by capillary action. When it passes the sample of mixture, the molecules in the mixture start to travel up the paper with the solvent. Non-polar substances move more quickly up the paper strip.

Thin-layer chromatography (TLC) works in a similar way to paper chromatography.

R_f values

The R_f (retention factor) value is a measure of the distance moved by a particular molecule:

$$R_f = \frac{\text{distance moved by molecule}}{\text{distance moved by solvent front}}$$

The R_f value is specific to the molecule, the solvent and the type of paper or thin layer used for separating the mixture. It can be used to identify a molecule in the mixture.

Revision activity

Write a list of all the key terms used in this chapter, then add the meaning of each key term.

Exam practice

1 (a) List three functions of carbohydrates in living things. [3]
(b) Describe and explain how the structure of starch (amylose) makes it suitable as a storage compound. [6]
(c) Complete the following table to compare glycogen with cellulose. [5]

Structural feature	Glycogen	Cellulose
Sugar(s) present		
Bonds present		
Branched or unbranched		
Coiled or straight		
Forms cross-links with other molecules		

2 (a) (i) Explain what is meant by the term *primary structure of a protein.* [1]
(ii) Explain what is meant by the term *secondary structure of a protein.* [3]
(b) Describe the bonds involved in holding the tertiary structure of a protein. [4]
3 (a) Explain why hydrogen bonds form between water molecules. [3]
(b) Explain the role of hydrogen bonds in making water a suitable medium for transport in living things. [10]
4 A student carried out a test to determine if a certain substance was present in the seeds of a small plant. She crushed the seeds and added some alcohol. She then filtered the solution before adding water to the filtrate. The resulting solution turned milky.
(a) State what substance was tested for. [1]
(b) Why was it necessary to filter the solution? [2]
(c) Suggest how the student could make this test quantitative. [6]

Answers and quick quiz 3 online

ONLINE

Summary

By the end of this chapter you should be able to:

- Describe how hydrogen bonding occurs.
- Relate the properties of water to its roles in living things.
- Describe the structure of α-glucose and β-glucose and the formation of glycosidic bonds.
- Describe and compare the structures of amylose, glycogen and cellulose and explain how the structure of each molecule relates to its function.
- Describe and compare the structures of a triglyceride and a phospholipid and explain how the structure of each molecule relates to its function.
- Describe the structure of amino acids and the formation of peptide bonds.
- Explain the four levels of protein structure with reference to the types of bonds involved at each level.
- Describe and compare the structures of haemoglobin, an enzyme and a hormone as examples of a globular proteins.
- Understand the properties and functions of fibrous proteins.
- Know the functions of a range of inorganic ions in living things.
- Describe the chemical tests for protein, reducing sugar, non-reducing sugar, starch and lipid.
- Describe how tests can be made quantitative.
- Understand the principles of chromatography.

4 Nucleotides and nucleic acids

The structure of nucleotides and nucleic acids

RNA and DNA

Ribonucleic acid (RNA) and **deoxyribonucleic acid (DNA)** are **nucleic acids** made from **nucleotides** (Figure 4.1). A nucleotide has three components:

- a phosphate group
- a pentose sugar
- an organic base

Table 4.1 outlines the differences between the nucleotides in RNA and those in DNA.

Table 4.1 The differences between the nucleotides in RNA and DNA

Feature	RNA	DNA
Pentose sugar	Ribose	Deoxyribose
Purines (two rings)	Adenine and guanine	Adenine and guanine
Pyrimidines (one ring)	Cytosine and uracil	Thymine and cytosine

Nucleic acids are complex organic substances present in living cells DNA or RNA molecules consist of many nucleotides linked in a long chain.

Nucleotides are **monomers** from which nucleic acids are formed. They are a combination of a phosphate, a pentose sugar and an organic base.

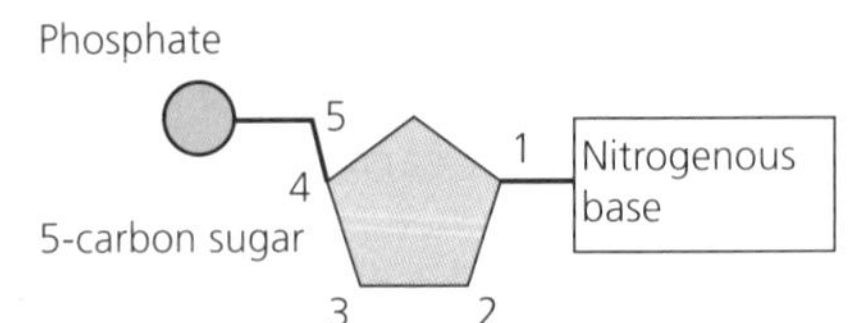

Figure 4.1 All nucleotides have the same structure. Note that the phosphate is attached to the fifth carbon atom in the sugar

The structure of RNA

RNA is a **polynucleotide**. It is usually single-stranded and much shorter than DNA. The sugar involved is ribose, not deoxyribose. The bases used include adenine, cytosine, guanine and uracil. There are three types of RNA:

1. **messenger RNA (mRNA)**, which carries the code held in the genes to the ribosomes where the code is used to manufacture proteins
2. **transfer RNA (tRNA)**, which transports amino acids to the ribosomes
3. **ribosomal RNA (rRNA)**, which makes up the ribosome

The structure of DNA

DNA is also a polynucleotide. The organic bases can pair up — one purine with one pyrimidine. They pair according to their complementary shapes. Adenine always pairs to thymine (or uracil in RNA) using two hydrogen bonds. Cytosine always pairs to guanine using three **hydrogen bonds**. The two polynucleotide strands lie in opposite directions, which is known as **antiparallel**. The two single polynucleotide strands are joined together to make a double strand. The whole molecule twists to form a helix, hence the name **'double helix'** (Figure 4.2).

Typical mistake

Candidates sometimes forget that the sugar in RNA is ribose not deoxyribose, as found in DNA.

Exam tip

Make sure that you know the key differences between RNA and DNA.

Revision activity

Draw a table to compare the structures of RNA and DNA.

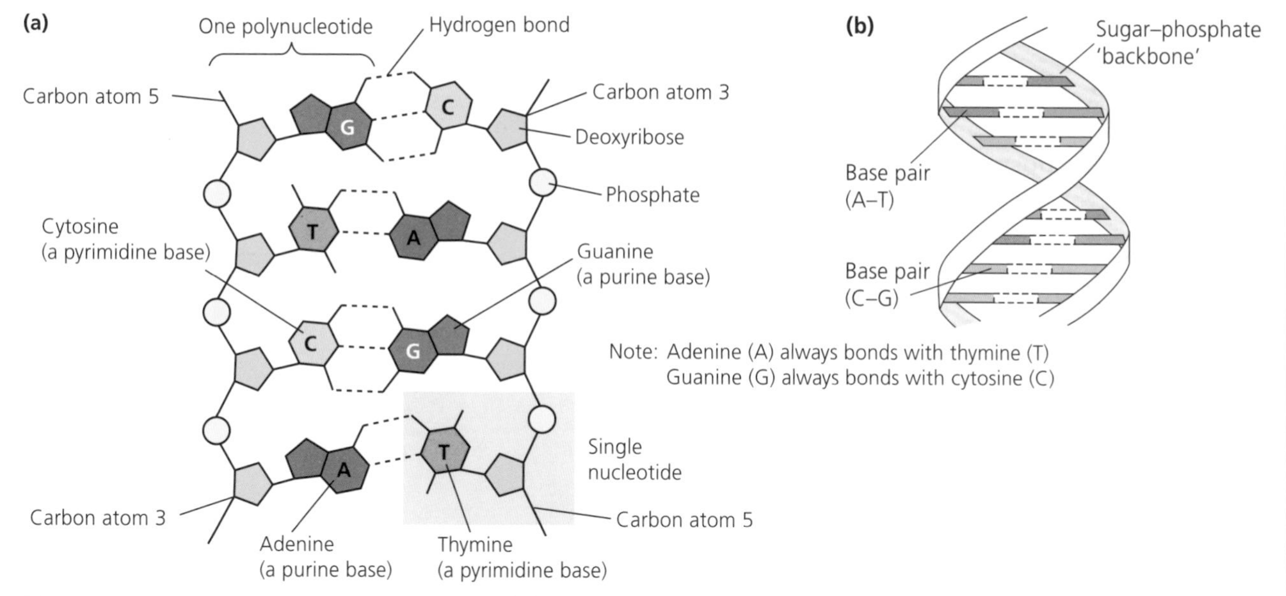

Figure 4.2 The structure of DNA showing (a) the hydrogen bonds and (b) the double helix

> **Typical mistake**
>
> Many candidates don't supply the detail about the number of hydrogen bonds used: adenine binds to thymine with two hydrogen bonds whereas cytosine bonds to guanine with three. This is why adenine cannot bind to cytosine and thymine cannot bind to guanine.

Polynucleotides

A polynucleotide is formed when nucleotides bind together in a long chain (Figure 4.3). The bonds are formed by condensation and are called **phosphodiester bonds** (Figure 4.4). They can be broken by hydrolysis. These bonds form between the sugar of one nucleotide and the phosphate group of another, making a sugar–phosphate 'backbone'. This leaves the organic base of each nucleotide sticking out to the side of the chain.

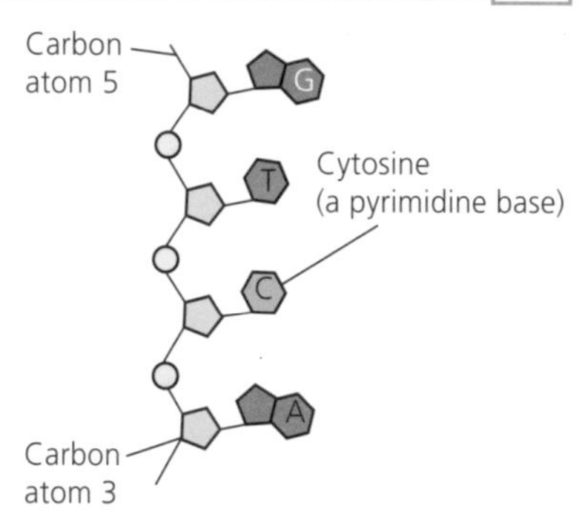

Figure 4.3 A polynucleotide

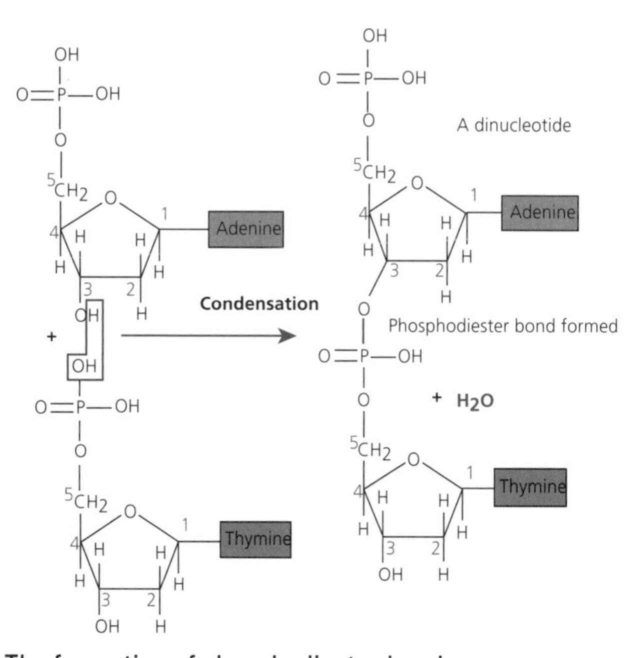

Figure 4.4 The formation of phosphodiester bonds

ADP and ATP

REVISED

ADP and **ATP** are **phosphorylated nucleotides**. They contain a **pentose sugar** (**ribose**), a **nitrogenous base** (**adenine**) and two or three **inorganic phosphates** (Figure 4.5).

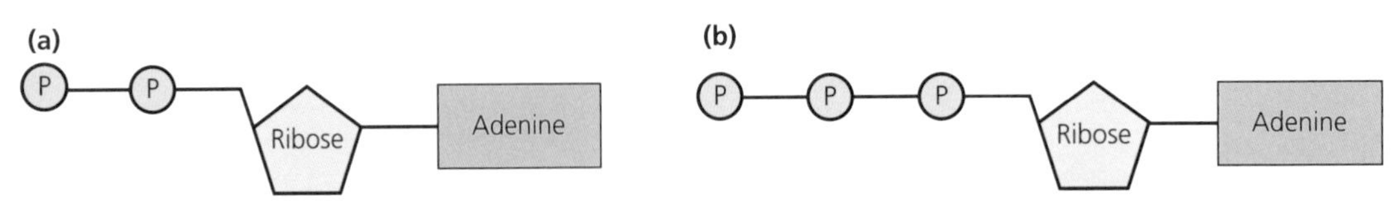

Figure 4.5 (a) ADP and (b) ATP are phosphorylated nucleotides

The replication of DNA

Semi-conservative DNA replication

REVISED

DNA **replication** occurs in all living organisms in order to copy their DNA for biological inheritance. Precise replication can take place because of the double-stranded structure of the DNA molecule. The process, known as **semi-conservative replication**, is as follows:

1. One double-stranded molecule untwists and the hydrogen bonds between the base pairs break. This is catalysed by the enzyme **helicase**.
2. The two polynucleotide chains separate, exposing the bases.
3. Each strand is then used as a template to make two new double strands.
4. New nucleotides pair to the exposed bases on both strands, using their complementary shapes to pair correctly.
5. Each new chain of nucleotides is bonded together by the enzyme **DNA polymerase** to form the second half of each DNA molecule.
6. The enzyme also checks that the pairing of the bases is correct.
7. Each new molecule then twists to form its double helix.

Replication means to make an identical copy.

Semi-conservative replication means that half of the original molecule is conserved in each of the new molecules.

Mutations

The precise replication of DNA is essential to ensure that identical copies of the genes are included in every cell of the body. Usually the precise pairing of the bases ensures that two exact copies of the original DNA molecule are made. However, occasionally an incorrect base may be bonded into place, which is known as a **mutation**. Mutations are **random** and **spontaneous**.

Revision activity

Draw a flow diagram to show the sequence of events in DNA replication.

Now test yourself

TESTED

1 Explain why it is important that each new double helix is identical to the original.

Answer on p. 223

The nature of the genetic code

The genetic code

REVISED

The **genetic code** is **universal** — all organisms use the same code. It is a sequence of bases. Three consecutive bases (known as **triplets**) codes for one amino acid. There are 64 possible triplet codes but only

20 amino acids used to make proteins. Some amino acids are coded for by more than one triplet (called **degenerate** codons). The triplets are **non-overlapping**. A particular sequence of triplets will code for one sequence of amino acids. A sequence of amino acids forms one polypeptide. The length of DNA that codes for one polypeptide is called a **gene**. One gene codes for one polypeptide. This may form a protein. If the protein contains more than one polypeptide (i.e. it has a quaternary structure), there will be more than one gene used to code for the protein.

Typical mistake

Many candidates confuse genes and DNA. DNA is a large molecule and a gene is a section of that larger molecule.

Now test yourself

2 Explain why the nucleotides are read in triplets.

Answer on p. 223

TESTED

The synthesis of polypeptides

The **synthesis of polypeptides** involves two stages:

1 **Transcription** — reading the code and producing a messenger molecule to carry the code out to the cytoplasm.
2 **Translation** — converting the code to a sequence of amino acids.

Transcription

REVISED

The genetic code is held in the nucleus. The DNA molecule is too large to leave the nucleus, so a smaller messenger molecule is made called messenger RNA (mRNA). The double-stranded DNA molecule is unwound and split by the action of the enzyme **RNA polymerase**, which breaks the hydrogen bonds holding the two strands together. This exposes the sequence of bases in the gene. The coding strand carries the genetic code and the complementary strand is a non-coding strand sometimes called the antisense strand or the template strand, which is used to build a copy of the coding strand called messenger RNA (mRNA). It has an identical sequence of bases to the coding strand of the gene, except that the thymine is replaced by uracil. Each triplet of bases on the mRNA is called a codon. The enzyme RNA polymerase joins the bases of the RNA to produce a complete single-stranded molecule. The base pairing is given in Table 4.2.

Transcription is the conversion of the genetic code to a sequence of nucleotides in mRNA. A codon is a triplet of bases in the mRNA.

Translation is the conversion of the code in mRNA to a sequence of amino acids.

The molecule of mRNA detaches from the DNA template and the DNA strands can rejoin to remake the double helix. The mRNA leaves the nucleus via the nuclear pores and enters the cytoplasm.

Table 4.2 Base pairing of the DNA template and mRNA

DNA template	mRNA
A	U
T	A
C	G
G	C

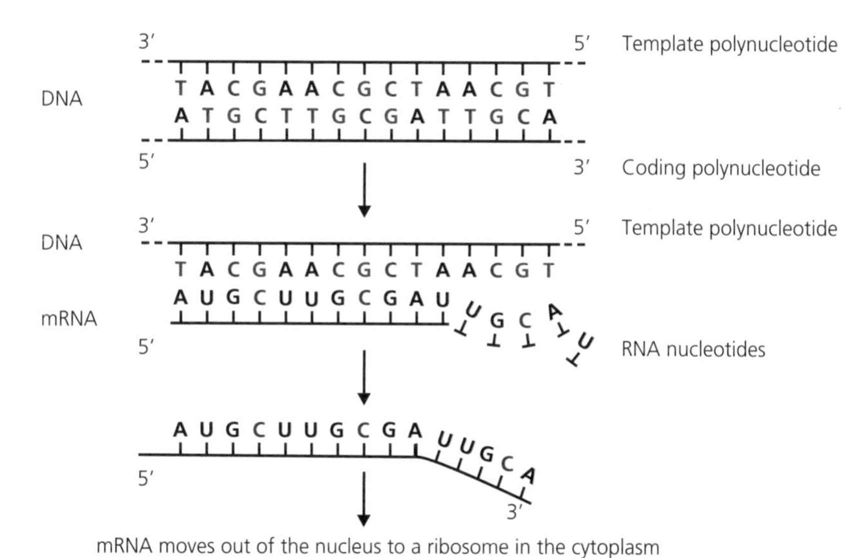

Figure 4.6 Transcription is the creation of a molecule of mRNA

Now test yourself

3 Explain why the genetic code must be converted to mRNA before it leaves the nucleus.

Answer on p. 223

TESTED

Translation

REVISED

Translation is the conversion of the genetic code to a sequence of amino acids (Figure 4.7). The mRNA joins on to a ribosome in the cytoplasm (ribosomes consist of ribosomal RNA associated with enzymes). The ribosome contains enough space for two codons at a time. The mRNA slides through the groove between the two components of the ribosome. As each codon enters the ribosome, it is used to position the next amino acid. Amino acids must be activated — they are combined with a specific molecule of transfer RNA (tRNA) that has a specific base triplet called the anticodon. The anticodon of the tRNA is complementary to a codon on the mRNA. ATP is used in this activation process. The amino acids attached to tRNA molecules are aligned with the correct part of the mRNA by the complementary pairing of the bases in the codon and anticodon. Enzymes bind the amino acids together in a chain by condensation reactions to create a growing polypeptide chain. The tRNA molecule is then released to be reused. When the ribosome reaches the end of the mRNA, the complete polypeptide chain is released and folds to form the secondary and tertiary structure of the protein. Some proteins need to be activated by cyclic AMP (cAMP), which interacts with the new protein to alter its three-dimensional shape. This makes the protein a better shape to fit their complementary molecules.

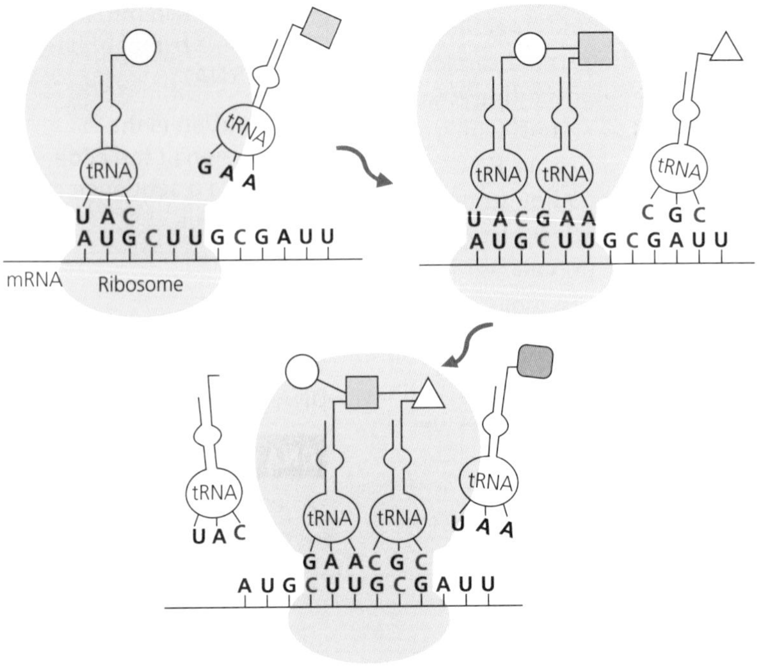

Revision activity

Draw a diagram of a cell including the organelles relevant to protein synthesis, then draw a flow diagram of protein synthesis underneath.

Exam tip

Ensure that you understand how the structure of DNA and the different forms of RNA enable them to perform their functions.

Revision activity

Write a list of all the key terms used in this chapter, then add the meaning of each key term.

Figure 4.7 Translation: amino acids are aligned according to the triplet sequence in the mRNA

Now test yourself

4 Explain why mRNA must be single-stranded.

Answer on p. 223

TESTED

Exam practice

1 (a) Describe the structure of a DNA nucleotide. [3]
(b) State two differences between the nucleotides of DNA and those of RNA. [2]
(c) Explain why RNA molecules are much smaller than DNA molecules. [2]

2 In a well-known experiment, scientists allowed bacteria to grow on a medium that contained only heavy nitrogen. The bacteria incorporated this nitrogen into their DNA. After many generations, the scientists assumed that all the nitrogen in the DNA had been replaced by heavy nitrogen. They then placed the bacteria on a medium containing normal nitrogen and allowed the bacteria to breed. The DNA was extracted from the bacteria after successive generations and separated according to its mass. This separation technique involved centrifuging the DNA in a dense solution, which caused

bands of DNA to appear in the centrifuge tube. Each band represents DNA of a different mass. Heavier DNA appears in a band lower down the tube. The results are shown in the diagram below.

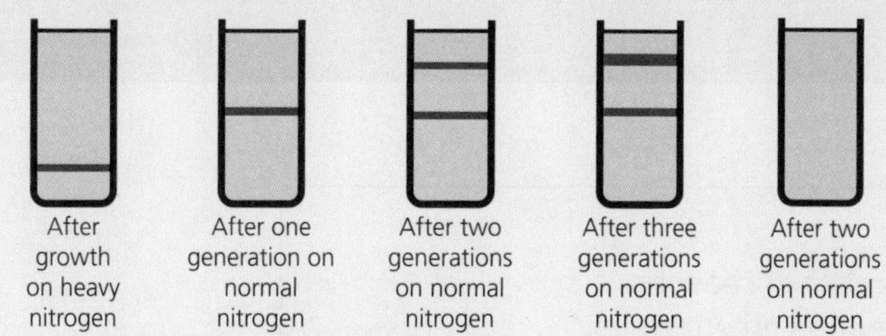

(a) Which part of the DNA contains nitrogen? [1]
(b) The scientists believed that this experiment demonstrated that DNA is replicated in a semi-conservative fashion. Explain what is meant by *semi-conservative replication*. [3]
(c) Explain how the results in the diagram demonstrate semi-conservative replication. [4]
(d) If replication had been conservative — where whole molecules of DNA are conserved and totally new ones are made — the results would show a different banding pattern. On the far right-hand diagram above, show what banding you would expect to see using DNA collected from bacteria after two generations on normal nitrogen. [2]

3 The following table shows information about base ratios in three species.

Species	Percentage composition of bases			
	A	C	G	T
X	28.2	21.0	20.8	29.2
Y	24.3	24.9		
Z	34.2	15.8		

(a) State the full names of the bases A, C, G and T. [4]
(b) Complete the table. [2]
(c) Suggest why the total for species X is not exactly 100%. [2]

Answers and quick quiz 4 online

ONLINE

Summary

By the end of this chapter you should be able to:

- Describe the structure of nucleotides as monomers.
- Describe the structure of RNA and how it differs from that of DNA.
- Describe the structure of DNA as a double-stranded polynucleotide with four bases called adenine, thymine, cytosine and guanine.
- Describe how the bases pair in a complementary fashion.
- Describe ADP and ATP as phosphorylated nucleotides.
- Outline the semi-conservative method of DNA replication.
- Describe the genetic code.
- Describe transcription and translation as stages in the synthesis of polypeptides.

5 Enzymes

Exam tip

This is an important topic for a number of reasons:

- It may be tested directly in a question on enzymes.
- It is likely to be tested as part of your practical work.
- It is the underlying principle for many topics in biology.

The role of enzymes

As intracellular and extracellular catalysts

REVISED

Enzymes are **globular proteins**. They act as **catalysts** to metabolic reactions in living organisms, which means they usually speed up metabolic reactions so that they occur at a reasonably fast pace even at body temperature.

Enzymes are required to build all the structures of the body (e.g. the cytoskeleton of a cell can be built up and reduced by enzyme activity), as well as to control the activity of the body.

Enzymes may be **intracellular** (working inside cells), such as **catalase** which converts hydrogen peroxide to oxygen and water. Alternatively, enzymes may be **extracellular** (working outside cells), such as the digestive enzymes **amylase** and **trypsin**, which are released into the digestive system.

A **globular protein** is a protein that folds and coils into a globular shape rather than a fibre.

Intracellular enzymes work inside cells.

Extracellular enzymes work outside cells.

The mechanism of enzyme action

Enzyme properties

REVISED

Enzymes have particular properties. These include:

- the molecule has a three-dimensional shape — its **tertiary structure**
- part of the molecule is an **active site** that is complementary to the shape of the substrate molecule
- each enzyme is specific to the substrate
- there is a high turnover number
- they have the ability to reduce the energy required for a reaction to occur
- their activity is affected by temperature, pH, enzyme concentration and substrate concentration
- the enzyme is left unchanged at the end of the reaction

Specificity and the lock and key hypothesis

The **specificity** of an enzyme refers to its ability to catalyse just one reaction or type of reaction. Only one particular substrate molecule will fit into the active site of the enzyme molecule. This is because of the shape of the active site.

The shape of the active site is caused by the specific sequence of amino acids. This produces a specific tertiary structure — the three-dimensional shape of the molecule. This is referred to as the **lock and key hypothesis** (Figure 5.1).

Typical mistake

Candidates tend to describe enzymes 'breaking down' the substrate. It is far more precise to say 'hydrolyse' or 'oxidise', i.e. to name the actual reaction that occurs.

Exam tip

Recalling a list of the properties of enzymes is easy. You need to be able to *explain* how those properties are related to the structure of the enzyme molecules.

The **lock and key hypothesis** explains how enzymes are specific to their substrate.

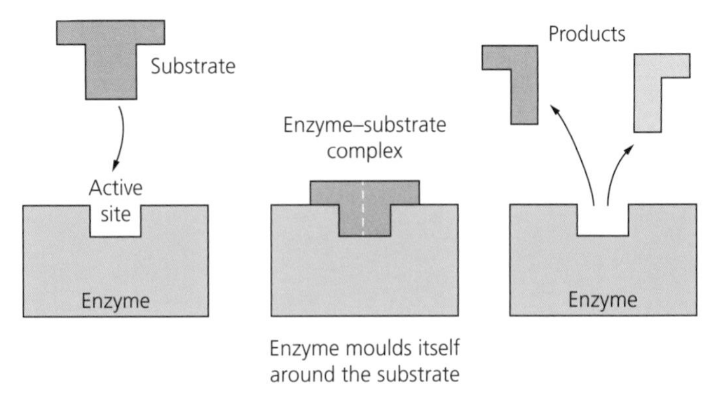

Figure 5.1 The enzyme has an active site that is complementary in shape to the substrate

Catalysing the reaction

Enzymes can speed up the rate of a reaction at body temperature. They lower the **activation energy** required for the reaction to occur. The activation energy is the amount of energy required to set off the reaction and break the bonds in the substrate molecule.

The induced-fit hypothesis

The **induced-fit hypothesis** helps to explain how the activation energy may be reduced.

The active site of an enzyme molecule does not have a perfectly complementary fit to the shape of the substrate. When the substrate moves into the active site, it interacts with the active site and interferes with the bonds that hold the shape of the active site. As a result, the shape of the active site is altered to give a perfect fit to the shape of the substrate. This changes the shape of the active site, which also affects the bonds in the substrate, making them easier to make or break (and therefore reducing the activation energy).

> **Typical mistake**
>
> Some candidates think that the lock and key hypothesis and the induced-fit hypothesis are mutually exclusive. However, the induced-fit hypothesis is a way to explain how a substrate moves into an active site that then changes to fit the substrate like a lock and key.

The course of an enzyme-controlled reaction

In an enzyme-controlled reaction, the enzyme (E) and substrate (S) molecules combine to form the **enzyme–substrate complex (ESC)**. The substrate is converted to the product, forming an **enzyme–product complex (ESP)**. The product is finally released and the enzyme is then free to take up another substrate molecule. This process is shown in Figure 5.2.

Figure 5.2 An enzyme-controlled reaction

> **Typical mistake**
>
> Candidates tend to refer to the lock and key hypothesis incorrectly. They make statements such as 'the enzyme works by the lock and key method', which is incorrect. The lock and key hypothesis explains how enzymes are specific to their substrate; it does not explain how they work.

Now test yourself

1 Using the analogy of a boulder in a hollow at the top of a hill, explain the role of an enzyme helping to overcome the activation energy in a reaction.

Answer on p. 223

TESTED

> The **induced-fit hypothesis** is a hypothesis that modifies the lock and key hypothesis.

Now test yourself

2 Describe what bonds and interactions could cause the enzyme to change shape to wrap around the substrate more closely.

Answer on p. 223

TESTED

> **Typical mistake**
>
> Many candidates forget to mention the enzyme–substrate complex.

Effects of conditions on enzymes

pH

REVISED

All enzymes have an optimum **pH** — the pH at which they work best. Therefore, they will not work as quickly at a pH outside their optimum range (Figure 5.3). This is because the hydrogen ions that cause acidity affect the interactions between *R* groups.

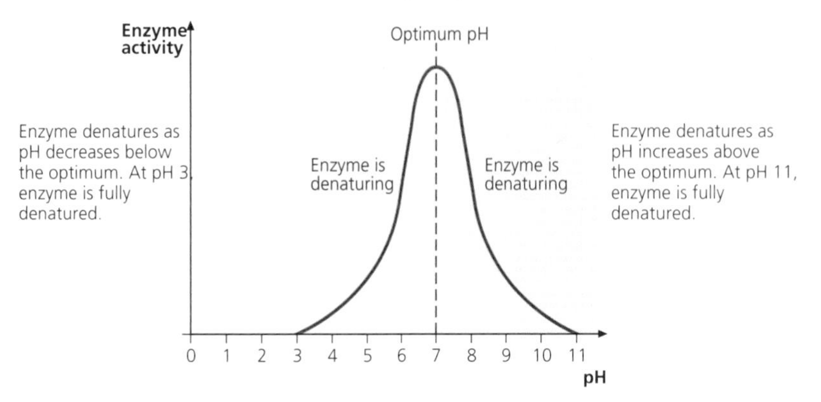

Figure 5.3 The effect of changing pH on enzyme activity

Altering the interactions between the *R* groups affects the tertiary structure of the molecule and may alter the shape of the active site. The shape will no longer be complementary to the shape of the substrate molecule.

Revision activity

Draw a graph to show the effect that pH change has on the activity of an enzyme. Annotate the graph with explanations for what happens at each stage.

Temperature

REVISED

The effects of **temperature change** on enzyme action vary depending on the temperature range considered (Figure 5.4). Each enzyme has an optimum temperature at which it is most active. This temperature is often 37°C (in mammals), but it may be different in other organisms.

At low temperatures (0–45°C), the activity of most enzymes increases as temperature rises. At low temperatures, the molecules have little kinetic energy. They collide infrequently with the substrate molecules and activity is reduced. As temperature rises, the molecules gain more kinetic energy. They collide more frequently with the substrate molecules and are more likely to have sufficient energy to overcome the required activation energy. Therefore, activity increases.

At higher temperatures, enzymes lose their shape (they become denatured). Higher temperatures cause increased vibration of parts of the molecule. If the temperature rises above a certain point, the bonds within the enzyme molecule vibrate too much and break, which alters the bonding in the active site, changing its shape. The active site no longer fits the shape of the substrate and activity reduces quickly to zero.

Typical mistake

Many candidates seem to think that all enzymes have the same optimum temperature and pH, which is not the case.

Revision activity

Draw a graph to show the effect that temperature change has on the activity of an enzyme. Annotate the graph with explanations for what happens at each stage.

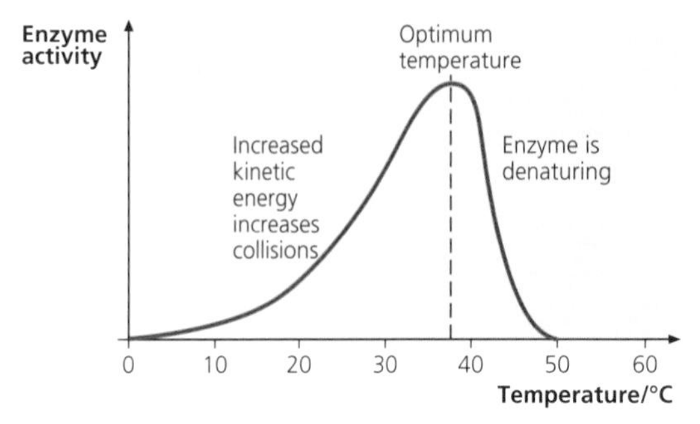

Figure 5.4 The effect of changing temperature on enzyme activity

Typical mistake

When describing or explaining the effects of changing conditions on enzyme action, many candidates make statements such as 'temperature affects enzyme activity'. They forget to say that *changing* the temperature affects enzyme activity.

Exam tip

To achieve full marks in the examination, you need to explain the effects of pH and temperature on the bonding in the active site and how this changes the shape of the active site. Remember that the changes are not instant — small increases above optimum will cause small changes and reduce activity. The enzyme does not suddenly stop working completely.

Enzyme concentration

REVISED

If there are more enzyme molecules in a particular volume of reaction medium, there are more active sites available. There is a greater likelihood of collisions between the enzyme and the substrate molecules. More interactions per second mean a higher rate of reaction. As the **enzyme concentration** increases, so does the rate of reaction (Figure 5.5).

The effect of pH and temperature on rates is usually because at extremes of pH and temperature some enzyme molecules are denatured and the concentration of active enzyme molecules is reduced.

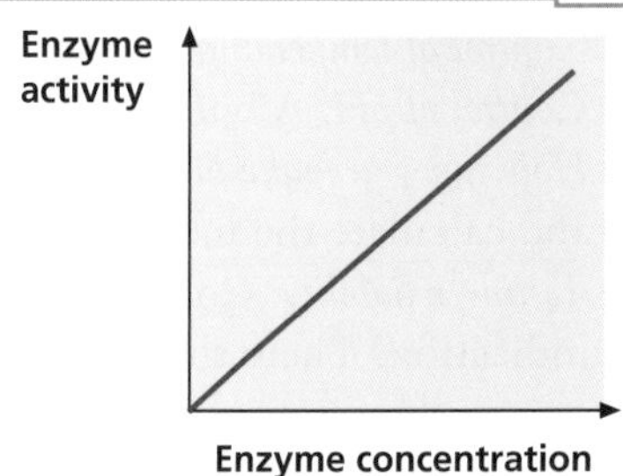

Figure 5.5 The effect of increasing enzyme concentration on enzyme activity

Substrate concentration

REVISED

If the **substrate concentration** is high, there is a greater chance of collisions between the enzyme and substrate molecules. As the substrate concentration increases, so does the rate of reaction (Figure 5.6).

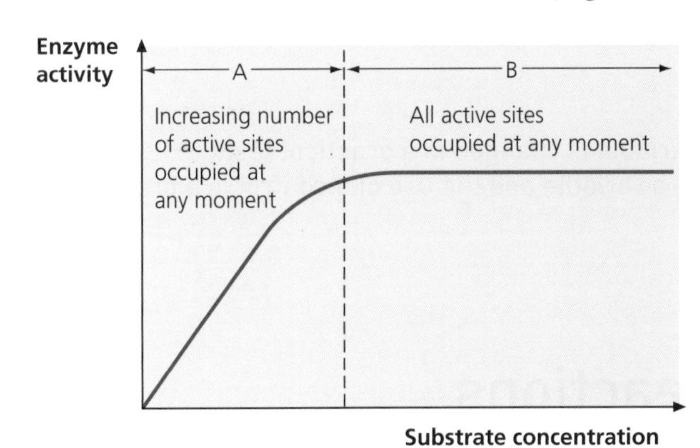

Figure 5.6 The effect of increasing substrate concentration on enzyme activity

If the number of enzyme molecules is limited, the rate of reaction levels off once all the enzyme active sites are fully occupied.

Exam tip

Questions could be limited to just one property, in which case it is likely that a lot of detail will be expected. Alternatively, this topic could be tested with a more open-ended type of question, in which case less detail will be expected on each property.

Revision activity

Draw a mind map to link together all the factors that affect enzyme activity and the explanations for each.

Practical investigations

Effects of pH, temperature, enzyme concentration and substrate concentration

REVISED

Exam tip

The written examination will include a number of marks to test your knowledge of practical work. These are often set in the context of a practical test carried out by a student or scientist. You may need to be able to describe a method, select suitable apparatus, evaluate a method, interpret results, analyse results or evaluate results.

You should be familiar with how the effects of changes in the following factors on enzyme activity can be investigated experimentally:

- pH
- temperature
- enzyme concentration
- substrate concentration

It is important to consider the following points.

1. *Volume and concentration of enzyme solution.*
2. *Volume and concentration of substrate solution.*
3. *Control of temperature.* A thermostatic water bath is often the best way.
4. *Control of pH.* A buffer solution controls pH.
5. *How can you make the test more reliable?* Repeat the test a number of times and calculate the mean.
6. *Testing reliability.* Comparing raw data to the mean is a good indication. There should be little variation around the mean — this can be shown using range bars. Calculating the standard deviation is even better.
7. *What is the control?* This is a test that omits one factor in the experiment to show that it is essential for the reaction to occur. It is usual to omit the enzyme from the reaction mixture to show that no reaction occurs without the enzyme.
8. *What level of precision is appropriate in the measurements?*
9. *How valid is the experiment?* Is it actually measuring what you think it should? Have you taken account of all possible conditions that may affect the reaction rate?

Exam tip

To gain full marks when describing a method, you must refer to the concentration and the volume of solutions.

Typical mistake

Many candidates refer to the amount of a substance when describing investigations. Ask yourself, what is the unit of 'amount'?

Revision activity

Draw the apparatus needed to collect and measure the volume of gas released from decomposing hydrogen peroxide. Annotate the diagram with information about how each variable can be controlled.

Now test yourself

TESTED

3. Explain why it is important that only one variable is changed in a practical test.
4. Explain the difference between controlling a variable and the use of a control in a practical test.

Answers on p. 223

Enzyme-controlled reactions

Coenzymes and cofactors

REVISED

Coenzymes are larger organic substances that take part in the reaction. They usually transfer other reactants between enzymes. Examples of coenzymes include coenzyme A, which takes part in aerobic respiration, and NAD, which is involved in transporting hydrogen atoms to the inner mitochondrial membrane. Coenzyme A and NAD are both made from the B vitamins.

Cofactors are inorganic substances, usually metal ions. They fit into the active site and activate the enzyme. Examples of cofactors include **Cl^-** in **amylase** and **Zn^{2+}** as a **prosthetic group** in **carbonic anhydrase**.

Coenzymes are large organic substances that are involved in some chains of enzyme-controlled reactions.

Cofactors are substances (usually a small metal ion) that are required to make enzymes function correctly.

Inhibitors

REVISED

Inhibitors are substances that reduce the rate of reaction and fit into a site on the enzyme.

Inhibitors are substances that reduce the activity of an enzyme.

Competitive inhibitors

Competitive inhibitors have a shape similar to the shape of the substrate. They fit into the active site, stopping the substrate molecules fitting in. This reduces the number of available active sites. The amount of inhibition depends on the relative concentrations of inhibitor and substrate molecules.

Non-competitive inhibitors

Non-competitive inhibitors fit into a different site on the enzyme molecule. They cause a change in the shape of the enzyme molecule. This affects the active site, so the substrate molecule can no longer fit in.

Reversible and non-reversible inhibitors

Reversible inhibitors occupy the enzyme site only briefly whereas **non-reversible inhibitors** bind permanently to the enzyme.

Drugs and poisons

Many **metabolic poisons** act by inhibiting enzymes. A poison such as cyanide inhibits the action of the enzyme cytochrome oxidase in aerobic respiration. Cytochrome oxidase contains iron ions and the cyanide binds to them. Many **medicinal drugs** also act as inhibitors of enzymes in the body. Aspirin binds to enzymes, preventing the formation of cell-signalling molecules that normally stimulate pain sensitivity. This is why taking the correct dose of medicinal drugs is important as overdosing can be lethal, especially if the inhibitor is non-reversible.

Product inhibition

Product inhibition occurs when the product of an enzyme-controlled reaction inhibits the enzyme. This can act to prevent too much product being formed.

Now test yourself

5 Explain how a non-competitive inhibitor prevents the substrate entering the active site.

Answer on p. 223

TESTED

Typical mistake

Many candidates think that competitive inhibitors are reversible and non-competitive inhibitors are irreversible, but this is not the case.

Revision activity

Write a list of all the key terms used in this chapter, then add the meaning of each key term.

Exam practice

1 Amylase is an enzyme that converts starch (amylose) to reducing sugars. A student investigated the effect of different temperatures on the rate of action of amylase. She prepared water baths containing starch suspension at four different temperatures. She then added amylase to each sample of starch and stirred. Each minute, the student transferred two drops of the starch/amylase mixture to a cavity tile containing iodine solution and noted the colour produced. The results are shown in the following table.

	Temperature (°C)			
Time (min)	**5**	**20**	**40**	**70**
1	Blue/black	Blue/black	Yellow	Blue/black
2	Blue/black	Blue/black	Yellow	Blue/black
3	Blue/black	Blue/black	Yellow	Blue/black
4	Blue/black	Dark yellow	Yellow	Blue/black
5	Blue/black	Yellow	Yellow	Blue/black
6	Blue/black	Yellow	Yellow	Blue/black
7	Blue/black	Yellow	Yellow	Blue/black

(a) (i) What does the colour blue/black indicate? [1]
(ii) What does the yellow colour in the samples at 20°C and 40°C indicate? [1]
(b) Suggest why the tube at 40°C was yellow after 1 minute. [2]
(c) Explain why the sample at 5°C remained blue/black. [3]
(d) Explain why the sample at 70°C remained blue/black. [4]
(e) As part of her evaluation, the student commented that it was difficult to sample all the tubes at the same time. Suggest one improvement she could make to her experiment. [1]
(f) She concluded that the optimum temperature for amylase activity is 40°C. Explain why this figure may not be accurate and suggest further improvements to her procedure that may make the test more accurate. [3]

2 (a) Enzymes are biological catalysts. Explain what is meant by the term *biological catalyst*. [2]
(b) Explain why more than one enzyme is needed to digest starch. In your response, ensure that the properties of enzymes are linked to their structure. [7]

3 (a) A student performed a practical procedure in which the rate of formation of maltose was measured in the presence and absence of chloride ions. In the presence of chloride ions, the rate of maltose formation increased.
(i) State the name given to a metal ion that increases the rate of an enzyme-controlled reaction. [1]
(ii) Suggest how the chloride ions act to have this effect on the rate of reaction. [2]
(b) The student extended his investigation to test the effect of temperature on the rate of reaction. When explaining his results, he made the following statement:

As the heat increased, the reaction went faster until it got to its highest. After this, the rate of reaction fell. This happened because the enzyme was killed and the hydrogen peroxide could not fit into the enzyme's key site.

Suggest a more appropriate word to replace each of the underlined words. [4]

Answers and quick quiz 5 online

ONLINE

Summary

By the end of this chapter you should be able to:

- State that enzymes are globular proteins with a specific tertiary structure, which catalyse metabolic reactions.
- State that enzyme action may be intracellular or extracellular.
- Describe the mechanism of action of enzyme molecules, with reference to specificity, active site, lock and key hypothesis, induced-fit hypothesis, enzyme–substrate complex, enzyme–product complex and lowering of activation energy.
- Describe and explain the effects of changes in pH, temperature, enzyme concentration and substrate concentration on enzyme activity.
- Explain the effects of competitive and non-competitive inhibitors on the rate of enzyme-controlled reactions, with reference to both reversible and non-reversible inhibitors.
- Explain the importance of cofactors and coenzymes in enzyme-controlled reactions.
- State that metabolic poisons and medicinal drugs may be enzyme inhibitors.

6 Biological membranes

Roles of membranes in cells

The plasma (cell-surface) membrane

REVISED

The plasma membrane is a **partially permeable barrier** between the cell and its environment. It keeps the contents of the cell separate from its environment, and limits what molecules can enter and leave the cell. It acts as the site for certain chemical reactions and enables the cell to communicate with other cells through the process of **cell signalling**.

Exam tip

Always remember to refer to the outer membrane of a cell as a plasma membrane or cell-surface membrane.

Revision activity

Draw a diagram of a cell to show all the membranes inside the cell-surface membrane.

Internal membranes (organelle membranes)

REVISED

Other membranes in the cell separate the **organelles** from the **cytoplasm**. These compartmentalise the cell, separating processes so that each process can occur in a specialised area of the cell. For example, all the enzymes involved in one process can be kept together and other processes do not interfere. Concentration gradients can be formed across the membranes. The membranes may act as the sites of specific **chemical reactions**, such as oxidative phosphorylation in aerobic respiration.

Revision activity

From memory, write a list of four functions of cell membranes.

Now test yourself

TESTED

1 Explain how a concentration gradient can be built up.
2 Suggest why a compartmentalised cell is more efficient than one that is not compartmentalised.

Answers on p. 223

The fluid mosaic model of membrane structure

REVISED

The **fluid mosaic model** describes the molecular arrangement of the membranes in a cell (Figure 6.1). A fluid mosaic membrane consists of:

- a bilayer of phospholipid molecules
- cholesterol which regulates the fluidity of the membrane, making it more stable
- glycolipids and glycoproteins that function in cell signalling or cell attachment
- protein molecules that float in the phospholipid bilayer. Some proteins are partially held on the surface of the membrane — these are called extrinsic proteins. Others are embedded in the membrane — these are called intrinsic proteins. Some proteins float freely in the bilayer whereas others may be bound to other components in the membrane or to structures inside the cell

The **fluid mosaic model** is the accepted structure of membranes in a cell.

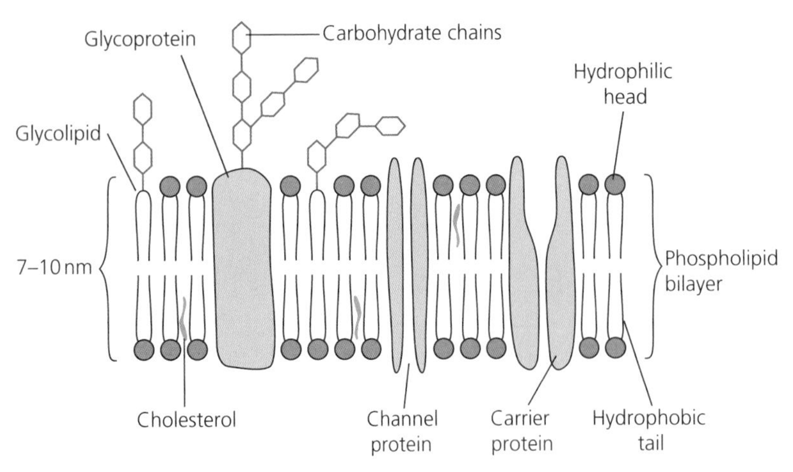

Figure 6.1 The fluid mosaic model of membrane structure

Phospholipids

Phosophopholipids form a barrier that limits movement of some substances into and out of the cell, or into and out of the organelles, so the membrane is partially permeable. Small, fat-soluble substances dissolve into the phospholipid bilayer and diffuse across the membrane. Water-soluble molecules and ions cannot easily dissolve and cross the membrane. Small molecules like water may diffuse across slowly, but most require special transport mechanisms.

Cholesterol

Cholesterol fits between the tails of the phospholipid molecules. It inhibits movement of the phospholipids, reducing the fluidity of the membrane. It also holds the phospholipid tails together, for mechanical stability. Cholesterol makes the membrane less permeable to water and ions.

Gycolipids and glycoproteins

The carbohydrate group on the protein or lipid molecule always has a specific shape and is used to recognise the cell — to identify it as 'self' or 'foreign'. Antigens on cell surfaces are usually **glycolipids** or **glycoproteins**.

Drugs and **hormones** can bind to these **membrane-bound receptors**. Medicines can be made to fit the receptors on certain cells. For example, asthmatics take salbutamol, which fits the receptors on smooth muscle in the airways to cause relaxation.

Cells communicate in an organism by cell signalling to coordinate the activities of the organism. The shape of the glycoprotein or glycolipid may be complementary to the shape of a signalling molecule in the body. Such complementary shapes can be used as binding sites to which the signalling molecules (e.g. hormones and neurotransmitter molecules in a synapse) attach. If the correct binding site is not present, the cell cannot respond to the signalling molecule. Binding sites are also used for cell attachment — the cells of a tissue bind together to hold the tissue together.

Proteins

Proteins have a variety of functions, such as enzymatic activity and cell signalling. However, many functions involve moving substances across the membrane. For example, some proteins may form:

- pores that allow the movement of molecules that cannot dissolve in the phospholipid bilayer
- carrier molecules that allow facilitated diffusion
- active pumps

Revision activity

From memory, draw a small diagram of a part of a cell-surface membrane to show the variety of molecules involved in its structure.

Exam tip

Never describe the cholesterol as increasing rigidity of the membrane.

Revision activity

Draw a diagram of a membrane including all the molecules involved in its structure and annotate it to describe what each component does.

Exam tip

Always say that the shape of the signal molecule is complementary to the shape of the receptor molecule.

Revision activity

Draw a table listing each membrane component in the left-hand column and its function in the right-hand column.

Now test yourself

3 Explain how the membrane can be selectively permeable.

Answer on p. 223

TESTED

Membrane structure and permeability

Temperature

REVISED

Membranes are partially permeable, fluid and stable at normal body **temperature**. If temperature increases, the molecules gain kinetic energy and move about more. This increases the permeability of the membranes to certain molecules. Any molecules that diffuse through the phospholipid bilayer will diffuse more quickly. This is because as the phospholipids move about, they leave temporary gaps between them, providing space for small molecules to enter the membrane.

If temperature increases further, the phospholipid bilayer may lose its mechanical stability (it may melt) and the membrane becomes even more permeable. Eventually, the proteins in the membrane will denature. This will further damage the structure of the membrane and it will become completely permeable.

Typical mistake

Many candidates describe the membrane as denaturing, but it is only the proteins that denature.

Revision activity

Describe how a cell-surface membrane may be different from the membranes inside a cell.

Solvents

REVISED

Solvents such as alcohol dissolve fatty substances. As the concentration of alcohol increases, the membrane is more likely to dissolve.

Now test yourself

4 Explain the importance of complementary shapes in cell signalling.

Answer on p. 223

TESTED

The movement of molecules across membranes

Passive transport

REVISED

Passive transport is the movement of molecules that does not need metabolic energy in the form of **adenosine triphosphate (ATP)**. It uses energy in the form of the kinetic (movement) energy. It only occurs when molecules move down a concentration gradient.

Because molecules move randomly, some may move in the 'wrong' direction — so you should describe passive transport as the net movement of molecules down their concentration gradient. Passive transport can occur in three forms:

- **Diffusion** (Figure 6.2) — the net movement of molecules away from a concentrated source. This may occur across a membrane if the molecules are fat-soluble or if they are small and can fit between the phospholipids in a membrane.
- **Facilitated diffusion** (Figure 6.2) — diffusion across a membrane that is helped by a **transport protein** in the membrane. The protein could be a pore protein (which may be gated) or it could be a carrier protein.
- **Osmosis** — the net movement of water molecules across a **partially permeable membrane**. Water molecules move down their **water potential gradient** (i.e. from an area of higher water potential to an area of lower water potential).

Passive transport is the movement of molecules without the use of metabolic energy.

Diffusion is the net movement of molecules down a concentration gradient.

Facilitated diffusion is diffusion that is aided by a protein in the membrane.

Transport proteins are proteins that help move substances across membranes. They include carrier proteins which may move molecules by changing shape.

Osmosis is the movement of water from a region of higher water potential to a region of lower water potential.

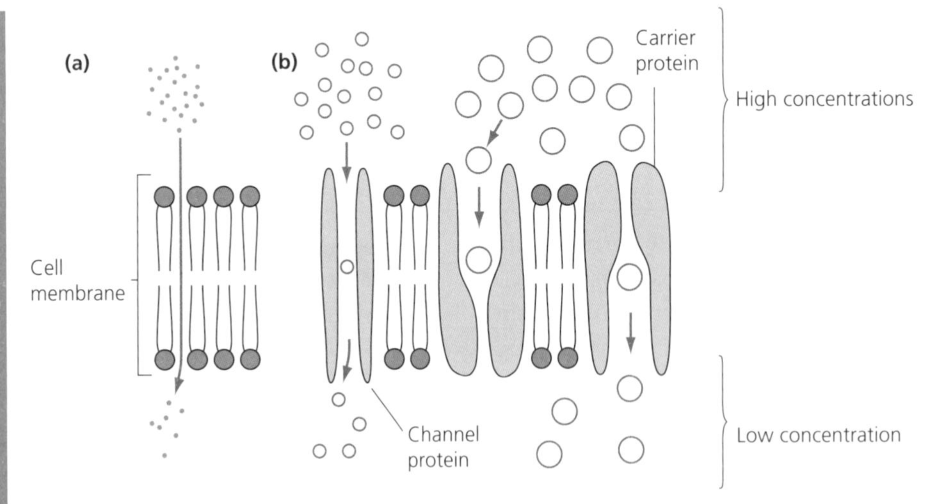

Figure 6.2 (a) Diffusion and (b) facilitated diffusion

The rate of diffusion

Diffusion occurs without using metabolic energy. It relies on the kinetic energy of the molecules. The rate of diffusion is affected by:

- Temperature — a higher temperature gives molecules more kinetic energy. At higher temperatures the molecules move faster, so the rate of diffusion increases.
- Concentration gradient — more molecules on one side of a membrane (or less on the other) increases the concentration gradient. This increases the rate of diffusion.
- Size of molecule — small molecules or ions can move more quickly than larger ones. Therefore, they diffuse more quickly than larger ones.
- Thickness of membrane — a thick barrier creates a longer pathway for diffusion, so diffusion is slowed down by a thick membrane.
- Surface area — diffusion across membranes occurs more rapidly if there is a greater surface area.

Exam tip

Don't describe or explain osmosis in terms of 'water concentration' as this can be confused with the concentration of solutes. Always use the term *water potential*.

Now test yourself

5 Explain why charged ions must be transported by facilitated diffusion rather than by simple diffusion.

Answer on p. 223

TESTED

Active transport

REVISED

Active transport (Figure 6.3) involves the movement of molecules using metabolic energy in the form of ATP. It can move molecules against their concentration gradient and uses membrane-bound proteins that change shape to move the molecules across the membrane.

Active transport is the movement of molecules against a concentration gradient using metabolic energy in the form of ATP.

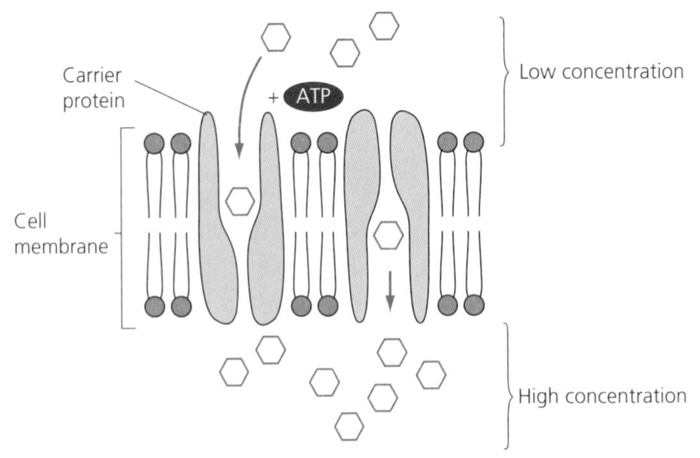

Figure 6.3 Active transport

Bulk transport

REVISED

Bulk transport is the movement of molecules through a membrane by the action of vesicles. **Endocytosis** is the formation of vesicles by the plasma membrane, which moves molecules into the cell. **Exocytosis** is the fusion of vesicles with the plasma membrane, which moves molecules out of the cell. Bulk transport uses metabolic energy.

Revision activity

Draw a mind map to show how a range of different substances pass through cell membranes.

Now test yourself

TESTED

6 Explain why proteins pass through membranes by bulk transport.

Answer on p. 223

Water potential and osmosis

Water potential

REVISED

Pure water has a **water potential** of zero. As solutes (sugars or salts) are added to a solution, the water potential gets lower. Therefore, a salt solution has a water potential below zero, i.e. a negative potential.

Water molecules will move from a solution with a higher water potential to a solution with a lower (more negative) water potential. Therefore, water molecules always move down their water potential gradient.

Water potential is a measure of the concentration of free water molecules.

Typical mistake

Some candidates confuse water potential and water potential gradient. A gradient can only exist between two places.

Osmosis

REVISED

A cell placed in water has a lower (more negative) water potential than the surrounding water. There is a water potential gradient from high outside the cell to lower inside the cell. As a result, water molecules enter the cell.

A cell placed in a strong salt solution has higher (less negative) water potential than the surrounding solution. There is a water potential gradient from higher inside the cell to lower outside the cell, so water molecules leave the cell.

Exam tip

It may be easier to describe osmosis in terms of water molecules moving from a less negative region to a more negative region.

Table 6.1 The effects of osmosis on animal and plant cells

	Solution	
Cell type	**Pure water**	**Strong salt**
Animal	An animal cell has no cell wall. The plasma membrane has no strength, so the animal cell will burst as water enters the cell.	An animal cell has no cell wall. The cytoplasm will shrink and the cell will shrivel. Its appearance is known as crenated.

Cell type	Solution	
	Pure water	Strong salt
Plant	Water makes a plant cell turgid. The vacuole is full of watery sap and the cytoplasm pushes the plasma membrane out against the cell wall. The cell wall is strong and will stop the cell bursting.	A plant cell will lose its turgidity. It will become flaccid. If the water loss continues, the cell vacuole will shrink. The cytoplasm will also shrink and the plasma membrane pulls away from the cell wall. This is called plasmolysis.

Now test yourself

TESTED

7 Explain why a plant cell will not burst when placed in pure water.
8 Using the terms *water potential* and *water potential gradient*, explain why a plant cell loses turgidity when placed in a strong salt solution.
9 In a plasmolysed plant cell, state what is found in the gap between the cell wall and the plasma membrane. Explain how it gets there.

Answers on pp. 223–24

Exam practice

1 **(a)** Describe how small molecules such as water and carbon dioxide can cross a plasma membrane. [2]
(b) Describe how large molecules can be brought into a cell. [3]

2 **(a)** The diagram below shows four animal cells that touch each other. Each cell has a different water potential as shown by the figures. Draw arrows to show the direction in which water will move by osmosis from cell to cell. [4]

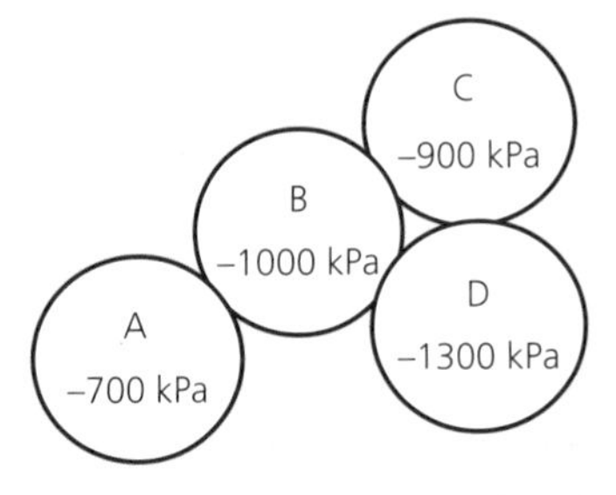

(b) Using the term *water potential*, explain the movement of water that would occur if cell C in the diagram were placed in pure water. [3]

3 Describe the functions of the following components of a cell membrane:
(a) cholesterol [1] **(b)** a glycoprotein [2] **(c)** phospholipids [2]

4 Use the idea of cell signalling to describe how the cell ensures that vesicles containing proteins can be transported to the correct organelle inside a cell. [4]

Answers and quick quiz 6 online

ONLINE

Summary

By the end of this chapter you should be able to:

- Outline the roles of membranes in cells.
- Describe the fluid mosaic model of membrane structure.
- Describe the roles of membrane components.
- Outline the effect of factors such as temperature and solvents on membrane permeability and structure.
- Understand how substances pass through membranes.
- Understand the terms *water potential* and *water potential gradient*.

7 Cell division, cell diversity and cellular organisation

The cell cycle

Processes during the cell cycle

REVISED

The cell cycle (Figure 7.1) is a series of events during which the cell duplicates its contents and splits in two. **Mitosis** is a small part of the cycle. The remainder of the cycle (known as **interphase**) is used for copying the **chromosomes** and checking the genetic information. During interphase, the cell also increases in size, produces new organelles and stores energy for another division.

Mitosis is the division of cells to produce two genetically identical daughter cells.

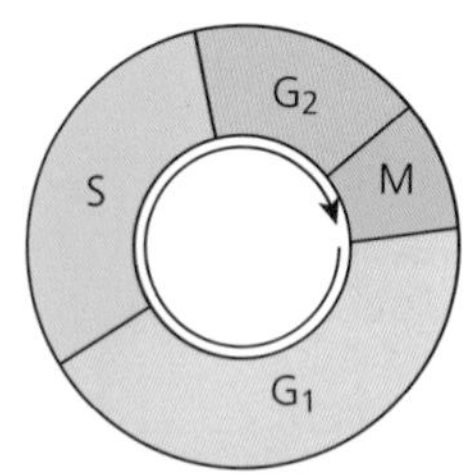

	Stage of cell cycle	Events within the cell	Control of cycle — checkpoints
INTERPHASE	G_1 phase (growth)	Cells grow and increase in size Transcription of genes (mRNA made) Protein synthesis occurs Organelles duplicate	G_1 cyclin-CDK complexes promote the production of transcription factors needed to produce enzymes for DNA replication G_1 checkpoint ensures that the cell is ready for DNA synthesis
INTERPHASE	S phase (synthesis)	DNA replicates, producing pairs of identical sister chromatids	Active S cyclin-CDK complexes ensure all DNA is replicated once
INTERPHASE	G_2 phase (growth)	Cells grow	G_2 checkpoint ensures that the cell is ready to enter M phase
	M phase (nuclear division or mitosis)	Cell growth stops Nuclear division consisting of stages: prophase, metaphase, anaphase and telophase Cytokinesis (cytoplasmic division)	Mitotic cyclin-CDK complexes promote the production of the spindle and the condensation of chromosomes Metaphase checkpoint ensures that the cell is ready to complete mitosis

Figure 7.1 The cell cycle

Now test yourself

TESTED

1 Explain why the cell must copy the DNA before mitosis.
2 Explain why the cell must produce new organelles and store energy during interphase.

Answers on p. 224

Mitosis

The significance of mitosis in life cycles

REVISED

Mitosis produces two genetically identical cells which are used for:
1 **growth** of the organism
2 **repair of tissues**

3 replacement of old cells
4 **asexual reproduction**

Asexual reproduction is the production of offspring from one parent and does not involve sex.

Typical mistake

Some candidates lose marks because they suggest that mitosis is for the growth or repair of cells. This is not true — cells grow during interphase and mitosis repairs tissues, not cells.

The main stages of mitosis

REVISED

Four stages make up mitosis in the cell cycle (Figure 7.2).

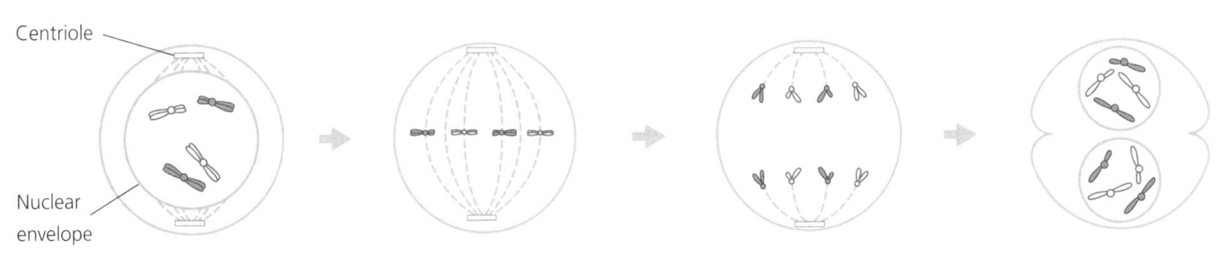

Prophase

- The start of mitosis
- Chromosomes shorten and thicken as DNA is tightly coiled
- Each chromosome is visible as two chromatids joined at the centromere
- Prophase ends as the nuclear envelope breaks into small pieces
- Centrioles organise fibrous proteins into the spindle

Metaphase

- Chromosomes are held on the spindle at the middle of the cell
- Each chromosome is attached to the spindle on either side of its centromere

Anaphase

- Chromatids break apart at the centromere and are moved to opposite ends of the cell by the spindle

Telophase

- Nuclear envelopes reform around the chromatids that have reached the two poles of the cell
- Each new nucleus has the same number of chromosomes as the original, parent cell
- The nuclei are genetically identical to each other

Figure 7.2 The stages of mitosis

Exam tip

Remember that the cell cycle and mitosis are continuous processes and the names of the stages reflect parts of a continuum.

Meiosis

The significance of meiosis in life cycles

REVISED

Meiosis is an alternative form of cell division. It produces four cells that are:

- not genetically identical
- gametes
- **haploid** (contain half the normal number of chromosomes)

The cells produced by meiosis are gametes — sex cells used for sexual reproduction. These cells contain one chromosome from each pair of **homologous chromosomes**. These have:

- the same shape and size
- the centromere in the same position
- the same genes in the same positions on the chromosomes

Homologous chromosomes are a pair of chromosomes that carry matching genes.

Typical mistake

Many candidates confuse meiosis with the process of fertilisation, which takes place after meiosis. In fertilisation, two gametes fuse to restore the diploid state in which the cell contains the homologous chromosomes.

The main stages of meiosis

REVISED

Before division starts (during interphase), the DNA replicates so that each chromosome consists of two identical copies called sister **chromatids** which are held together by a centromere. The cell divides twice and each division can be divided into four stages (Figure 7.3).

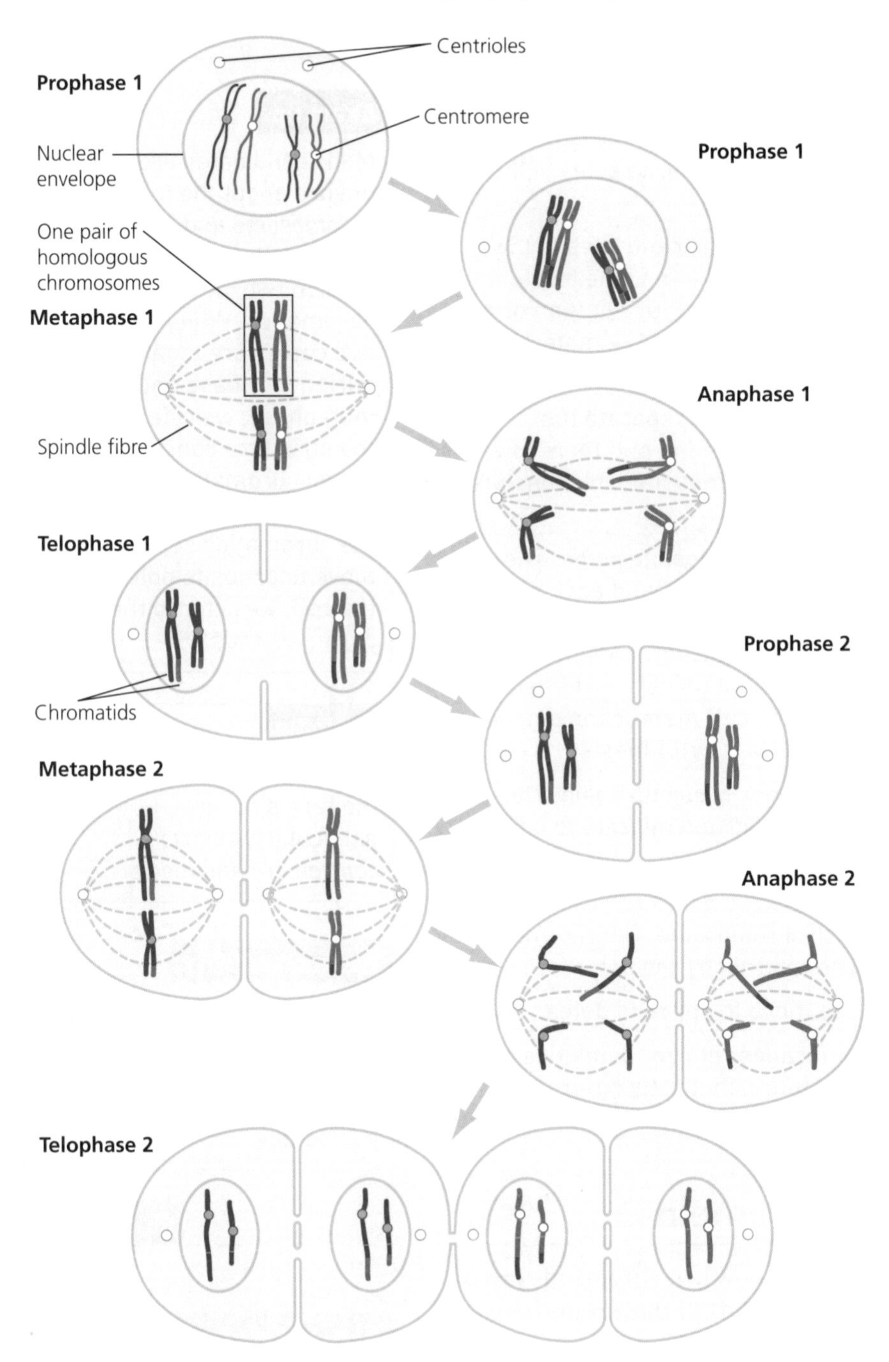

Figure 7.3 The stages of meiosis

Typical mistake

Many candidates confuse homologous chromosomes with sister chromatids. Remember that homologous chromosomes are the two chromosomes of similar length found in all diploid cells. They may contain identical alleles or alleles that are slightly different. Sister chromatids are the two identical copies of the same chromosome that are produced by replication of the DNA before cell division starts.

Table 7.1 The stages of meiosis

Stage of meiosis	Events within the cell
Prophase 1	Chromosomes condense — they supercoil to become shorter and thicker. Homologous chromosomes pair to form **bivalents** containing four chromatids. The chromatids in each bivalent break and rejoin to form **chiasmata** or crossovers — this is where sections of the non-sister chromatids can be exchanged. The nuclear membrane breaks up to form small membrane sacs. The **centriole** replicates and migrates to opposite poles of the cell and forms the **spindle fibres**.
Metaphase 1	Microtubules attach from the centrioles to the centromere of each chromosome. The bivalents move to the equator of the cell. Orientation of each bivalent on the equator is random — maternal or paternal chromosomes could be facing either pole.
Anaphase 1	The microtubules shorten to separate the homologous chromosomes and pull them towards opposite poles. Each chromosome still consists of two chromatids.
Telophase 1	The chromosomes reach opposite poles. The nuclear membrane reforms around each set of chromosomes to produce two nuclei. These nuclei are haploid as they have one chromosome from each homologous pair (but there are still two sister chromatids). The **cell membrane** pinches in to form two cells — this is **cytokinesis**.
Prophase 2	The nuclear membranes break up again. The centrioles replicate again and migrate to opposite poles of the two new cells.
Metaphase 2	Microtubules attach between the centrioles and the centromere of each chromosome. The chromosomes move to the equator and align randomly.
Anaphase2	Sister chromatids move to opposite poles.
Telophase 2	The nuclear membranes reform. Cytokinesis occurs to produce four genetically different haploid cells.

A **bivalent** is the pair of homologous chromosomes.

Chiasmata are the points at which chromatids cross over (singular: chiasma).

Exam tip

Many candidates are unsure about the terms *chromosome* and *chromatid*. A chromosome replicates to form two identical chromatids. While the two copies are together during prophase and metaphase, we refer to the structure consisting of two chromatids as a chromosome. However, once the chromatids separate and move to opposite poles of the cell, we refer to them as chromosomes.

Exam tip

Make sure you can spell the terms accurately, particularly *centriole*, *centromere* and *meiosis*.

Revision activity

Draw your own set of diagrams showing meiosis and annotate each sketch to show what is occurring at each stage.

Creating genetic variation in meiosis

REVISED

Genetic variation can be created by meiosis. When chromatids **cross over**, they exchange lengths of DNA (Figure 7.4). If this occurs between non-sister chromatids, it makes new combinations of alleles.

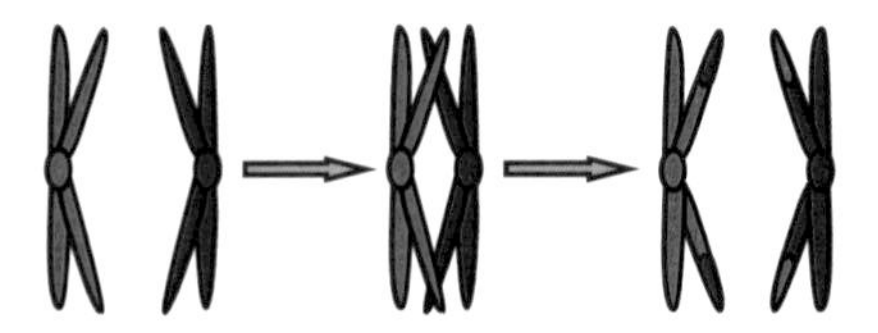

Figure 7.4 Exchanging genetic material between non-sister chromatids

Now test yourself

3 Explain why there must be two divisions in meiosis.

Answer on p. 224

TESTED

The way the bivalents orientate on the equator during metaphase 1 is random. This means that either the maternal or the paternal chromosome of a bivalent my face either pole. Therefore, the combination of maternal and paternal chromosomes migrating to either pole is random. This is called **independent assortment** of homologous chromosomes (Figure 7.5). In a similar way, the orientation of the chromosomes on the equator in metaphase 2 is random. Therefore, the combination of chromatids migrating to each pole is random. This is called independent assortment of sister chromatids.

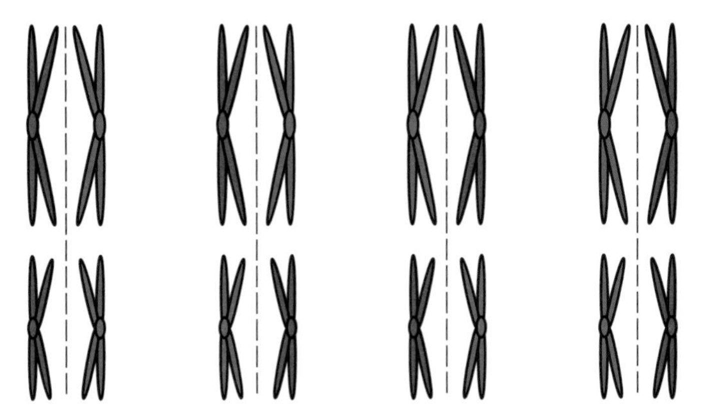

Figure 7.5 **Independent orientation of homologous chromosomes on the equator creates genetic diversity by independent assortment**

Cell specialisation in multicellular organisms

Cell specialisation for particular functions

REVISED

Cells of **multicellular organisms** are **specialised** in a number of ways, as shown in Table 7.2.

Squamous means flattened.

Table 7.2 **Cell specialisation**

Cell type and function	Specialisation	How that specialisation helps its function
Erythrocytes (red blood cells) carry oxygen in the blood	Small and flexible	To fit through tiny capillaries
	Full of haemoglobin	To bind to the oxygen
	No nucleus	To allow more space for haemoglobin
	Biconcave shape	To provide a large surface area to take up oxygen quickly
Neutrophils engulf and digest foreign matter or old cells	Flexible shape	To enable movement through tissues
	Lobed nucleus	To help movement through membranes
	Many ribosomes	To manufacture digestive enzymes
	Many lysosomes	To hold digestive enzymes
	Many mitochondria	To release the energy needed for activity
	Well-developed cytoskeleton	To enable movement
	Membrane-bound receptors	To recognise materials that need to be destroyed

Cell type and function	Specialisation	How that specialisation helps its function
Sperm carry the paternal chromosomes to the egg	Tail (flagellum)	To enable rapid movement
	Acrosome	To help digest egg surface
	Small	To make movement easier
	Many mitochondria	To release the energy needed for rapid movement
Epithelial cells act as surfaces	Often flat (**squamous**)	To cover a large area
	Often thin (squamous)	To provide a short diffusion distance
	May be ciliated	To move mucus
	May be cuboid	To provide a barrier
	Many glycolipids and glycoproteins in cell-surface membrane	To hold cells together or for cell signalling
Palisade cells for photosynthesis	Elongate	To fit many chloroplasts into the space
	Contain many chloroplasts	To absorb as much light as possible
	Show cytoplasmic streaming	To move the chloroplasts around
	Contain starch grains	To store products of photosynthesis
Root hair cells absorb water and mineral ions from the soil	Long extension (hair)	To increase surface area
	Active pumps in cell-surface membrane	To absorb mineral ions by active transport
	Thin cell wall	To reduce barrier to movement of ions and water
Guard cells control the stomatal opening	Active pumps in cell-surface membrane	To move mineral ions in and out of cell to alter the water potential
	Unevenly thickened wall	To cause the cell to change shape as it becomes more turgid
	Large vacuole	To take up water and expand to open the stoma

Revision activity

Draw a detailed diagram of each cell type in Table 7.2 and annotate it to explain how the cell is specialised.

Now test yourself

4 Explain how cells differentiate.

Answer on p. 224

TESTED

Tissues, organs and organ systems

Tissues

REVISED

A **tissue** is a collection of cells that work together to perform a particular function. They may be similar to each other or they may perform slightly different roles. For example:

- **Squamous epithelium** is a layer of flattened cells bound together to produce a surface.
  - **Ciliated epithelium** contains ciliated cells that move mucus over their surface and goblet cells that produce the mucus.

- **Cartilage** consists of cells called chondrocytes that secrete a matrix of collagen.
- Muscle is found in three types: smooth muscle consists of single cells that can contract; skeletal muscle forms multinucleate fibres containing protein filaments that slide past one another; cardiac muscle forms cross-bridges to ensure that the muscle contracts in a squeezing action.
  - **Xylem** contains vessels that carry water and xylem fibres for support.
  - **Phloem** contains two types of cell:
    - Sieve tube elements which form sieve tubes
    - Companion cells.

Now test yourself

5 Explain why forming tissues is more efficient than using individual cells to perform a task.

Answer on p. 224

TESTED

Organs

REVISED

An **organ** is a collection of tissues working together to perform a common function.

Organ systems

REVISED

An **organ system** is made up of two or more organs working together to perform a life function such as excretion or transport.

Revision activity

Write separate lists of all the tissues and organs in the human transport and gaseous exchange systems.

The features and differentiation of stem cells

Stem cells and differentiation

REVISED

Stem cells are cells that are not specialised or differentiated. They maintain the capacity to undergo mitosis and differentiate into a range of cell types. Differentiation is the ability of a cell to specialise to form a particular type of cell.

Stem cells are a **renewing source** of **undifferentiated cells** for the growth and repair of tissues and organs. During growth and repair, stem cells divide to produce new cells, which then differentiate to become specialised to their function. Stem cells have the ability to use all their genes. Differentiation occurs by switching on or off appropriate genes.

The production of blood cells

REVISED

Stem cells in the bone marrow divide and differentiate to form both red and white blood cells.

Erythrocytes (red blood cells)

Erythrocytes are specialised to carry oxygen. See Table 7.2 for specialisations. They have no nucleus and very few organelles, providing more space for haemoglobin molecules which are synthesised during development before the other organelles are lost.

Neutrophils (white blood cells)

Neutrophils are the most common type of phagocyte used to ingest and destroy bacteria. See Table 7.2 for specialisations.

The production of xylem vessels and phloem sieve tubes

REVISED

Xylem and phloem are transport tissues in plants. New cells are produced by mitosis in the **meristem**. These cells are expanded by the uptake of water and the development of a vacuole. They then differentiate into xylem and phloem.

Xylem

Lignin is deposited in their cell walls to strengthen and waterproof the wall. The cells die and the contents are removed as the end walls break down, forming continuous columns of cells. These form tubes with wide lumens to carry water and dissolved minerals. The lignification is incomplete in some places, forming bordered pits.

Phloem

Phloem consists of two types of cell that work together:

1. Sieve tube elements lose their nucleus and most of their organelles. The end walls develop numerous sieve pores to form sieve plates between the elements.
2. Companion cells retain their organelles and can carry out metabolism to obtain and use ATP to actively load sugars into the sieve tubes.

Sieve tube elements and companion cells are linked by numerous **plasmodesmata**.

Plasmodesmata (singular: plasmodesma) are connections between cells where the cytoplasm is continuous.

The potential uses of stem cells in research and medicine

REVISED

Stem cells can be sourced from different tissues:

1. embryonic stem cells, which are present in a young embryo
2. blood from the umbilical cord
3. adult stem cells found in developed tissues such as **bone marrow** (but scientists are finding stem cells in almost all tissues)
4. scientists can also induce certain tissue cells to become stem cells (known as induced pluripotent stem cells or iPS cells).

Stem cells can be used in the following ways in research and medicine.

Repair of damaged tissues

Stem cells have been used to treat certain conditions and continued research suggests that many uses can be found:

1. Stem cells in bone marrow are used to treat diseases of the blood, such as leukaemia.
2. Stem cells have been used to repair the spinal cord of rats.
3. Stem cells have been used to treat mice with type 1 diabetes.
4. Stem cells in the retina can be made to produce new light-sensitive cells.
5. Stem cells directed to become nerve tissue could be used to treat **neurological conditions** such as **Alzheimer's disease** and **Parkinson's disease**.
6. Stem cells may also be used to treat other conditions such as arthritis, stroke, burns, blindness, deafness and heart disease.

Developmental biology

Scientists use stem cells to gain a better understanding of how multicellular organisms develop, grow and mature.

1 They study how differentiation occurs — how cells develop to make particular cell types.
2 They study what happens when differentiation goes wrong.
3 They are trying to find out if they can re-enable differentiation and growth in adult cells to help tissue repair (healing) in later life or even the ability to re-grow an organ or limb.

Revision activity

Write a list of all the key terms used in this chapter, then add the meaning of each key term.

Exam practice

1 The following statements are about meiosis.
   - A Chromatids pair during prophase 1.
   - B Bivalents form during prophase 2.
   - C Homologous chromosomes are independently assorted.
   - D Four haploid cells are produced.
   - E Meiosis is used for growth and repair.

   Which of the following options identifies the correct statements? [1]
   - (a) All five statements are correct.
   - (b) Statements B, C and D are correct.
   - (c) Statements C and D are correct.
   - (d) All statements are incorrect.

2 (a) Explain what is meant by the term *differentiation*? [2]
   (b) Beta cells in the pancreas are specialised to produce the protein hormone insulin. Suggest how these cells may be specialised. [3]
   (c) Diabetes is a disease in which the beta cells stop producing insulin. Suggest how stem cells could be used to cure this disease. [2]
   (d) Explain the advantages of using stem cells from an embryo. [2]

3 (a) Define the terms *tissue* and *organ*. [4]
   (b) Plant transport tissues are called xylem and phloem. Describe how cells are organised to form these tissues. [5]

Answers and quick quiz 7 online

ONLINE

Summary

By the end of this chapter you should be able to:

- Describe the cell cycle and the stages of mitosis.
- Explain the significance of mitosis for growth, repair and asexual reproduction.
- Describe the stages of meiosis and explain the significance of meiosis in producing haploid cells.
- Understand the meanings of the terms *diploid, differentiation, haploid, homologous chromosomes, stem cell, tissue, organ* and *organ system*.
- Describe and explain how certain cells and tissues are specialised for their function.
- Explain how stem cells can be used in research and medicine.

8 Exchange surfaces

The need for specialised exchange surfaces

Surface area to volume ratio

REVISED

A living organism needs to absorb substances from its surroundings and remove waste products. However, as an organism increases in size, its volume increases so it needs more from its environment. Unfortunately, its surface area does not increase as quickly as its volume, so the larger an organism gets, the more difficult it becomes to absorb enough substances over its surface.

Table 8.1 shows what happens to surface area, volume and **surface area to volume ratio (SA : V)** as an organism increases in size.

The **surface area to volume ratio (SA : V)** is the surface area of an organism divided by its volume. It is a key concept as the surface area must be able to provide sufficient oxygen through diffusion from the environment.

Table 8.1 The effects on an organism as it increases in size

Length of organism (l) (mm)	Surface area of organism ($6 \times l^2$) (mm^2)	Volume of organism (l^3) (mm^3)	Surface area to volume ratio
1	6	1	6 : 1
5	150	125	1.2 : 1
10	600	1000	0.6 : 1

We can see that as size increases:

- surface area increases
- volume increases, but more quickly than surface area
- surface area to volume ratio decreases

Revision activity

Draw a graph of surface area to volume ratio plotted against body size for a range of body lengths from 5 μm (bacteria) up to 5 m (small whale). Remember to convert the units appropriately. Mark the graph to show where an amoeba, a mouse, a man and an elephant would fit on the graph.

The significance of surface area to volume ratio

Single-celled organisms are small and have a large surface area to volume ratio. Their surface area is large enough for sufficient oxygen and nutrients to diffuse into the cell to provide all its needs, and for wastes to diffuse out.

Multicellular organisms have a smaller surface area to volume ratio. Diffusion is too slow for the oxygen and nutrients to diffuse across the whole organism. The surface area is no longer large enough to supply all the needs of the larger volume. Therefore, a specialised exchange surface is required, such as the lungs which are used for gaseous exchange. Being multicellular and **metabolically active** also increases the need for a specialised exchange surface.

Exam tip

Remember that:

- length, surface area and volume all have different units
- surface area to volume ratio has no units
- suitable units must always be included in any work involving figures

Typical mistake

Many candidates confuse surface area with surface area to volume ratio. An elephant has a large surface area, but it has a small surface area to volume ratio.

Now test yourself

TESTED

1 List the factors that affect the need for a specialised surface for gaseous exchange.
2 Explain why a single-celled organism such as an amoeba does not need a specialised surface for gaseous exchange whereas a large tree does.

Answers on p. 224

Efficient exchange surfaces

The features of good gaseous exchange surfaces

REVISED

A good gaseous exchange surface (Figure 8.1) has certain features, as shown in Table 8.2.

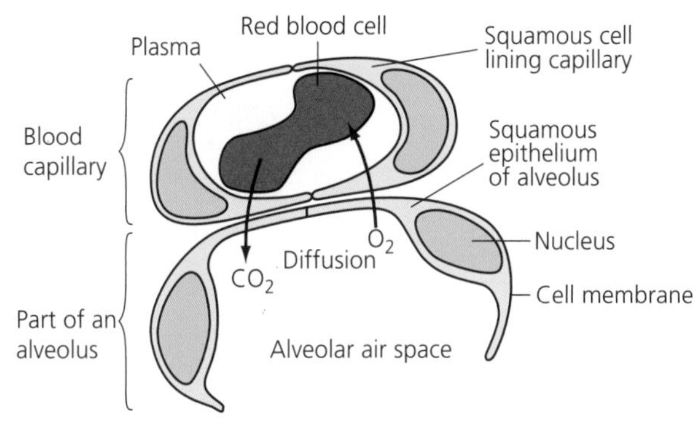

Figure 8.1 The gaseous exchange surface in the lungs

Table 8.2 The features of good surfaces for gaseous exchange

Feature	Reason	In the lungs
Large surface area	To provide space for molecules of oxygen and carbon dioxide to pass	Lung epithelium folded to form numerous alveoli (singular: alveolus)
Thin layer	To provide a short diffusion pathway	Lung epithelium and capillary endothelium are made from squamous cells
Steep concentration gradient	To ensure molecules diffuse rapidly in the correct direction	Good supply of blood on one side and ventilation of the air sacs on the other side

Steep concentration gradient

A steep **concentration gradient** is maintained by increasing the concentration of molecules on the supply side and reducing the concentration on the demand side. In the lungs, this is achieved by good blood flow and ventilating the air spaces. Blood flow brings carbon dioxide to the lungs and removes oxygen; ventilation brings oxygen to the lung surface and removes carbon dioxide.

Typical mistake

Many candidates describe the lungs as having a 'thin cell wall' — they probably mean a 'wall of thin cells' or a 'thin wall of cells'. This sort of vague wording should be avoided. Describe the barrier as creating a short diffusion pathway because the cells are squamous.

Revision activity

Draw a mind map to link the features of a good gaseous exchange surface to rate of diffusion.

A **concentration gradient** is the difference in concentration between two points.

Exam tip

Remember to describe changes in the concentration of the gases in the blood or in the air sacs as it is the concentration gradient that drives diffusion.

Now test yourself

TESTED

3 List the factors that affect the concentration gradient.

Answer on p. 224

Ventilation and gaseous exchange in mammals

The lungs

REVISED

Figure 8.2 shows the structure of the human gaseous exchange system. Table 8.3 summarises the distribution and function of lung cells and tissues.

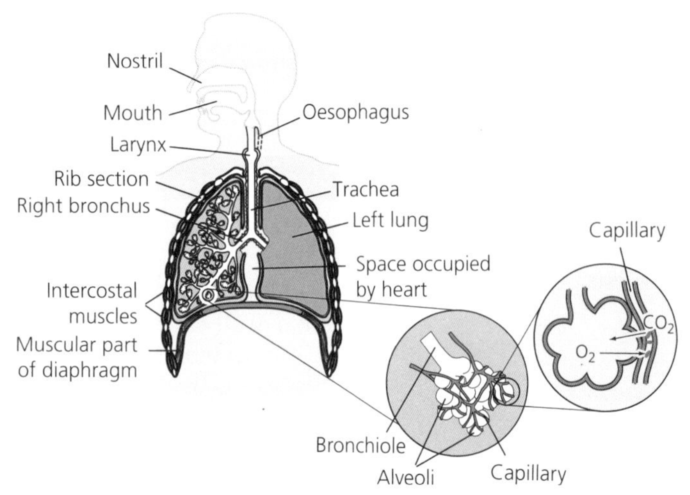

Figure 8.2 The human gaseous exchange system with details of the alveoli

Revision activity

From memory, write a list of the tissues in the lungs and describe what each does to help gaseous exchange.

Table 8.3 The distribution and function of cells and tissues in the lungs

Structure	Distribution	Function
Capillaries	Over surface of alveoli	To provide a large surface area for exchange
Cartilage	In walls of bronchi and trachea	To hold the airways open
Ciliated epithelium	On surface of airways	The cilia move or waft the mucus along
Elastic fibres	In walls of airways and over alveoli	To recoil to return the airway or alveolus to original shape. In alveolus this helps to expel air
Goblet cells	In ciliated epithelium	To produce and release mucus
Smooth muscle	In walls of airways	Contracts to constrict or narrow the airways
Squamous endothelium	Capillary wall	To provide a thin barrier to exchange — a short diffusion pathway
Squamous epithelium	Surface of alveoli	To provide a thin barrier to exchange — a short diffusion pathway

Figures 8.3 and 8.4 show details of the wall of the trachea and the distribution of tissues in the lungs.

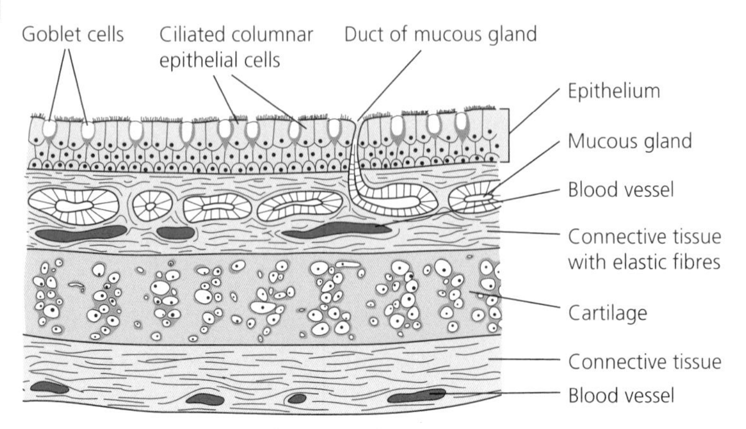

Figure 8.3 The wall of the trachea

Typical mistake

Many candidates make the mistake of describing:

- cilia as 'hairs'
- cilia as trapping bacteria
- elastic tissue as contracting
- smooth muscle as providing a smooth surface to reduce friction

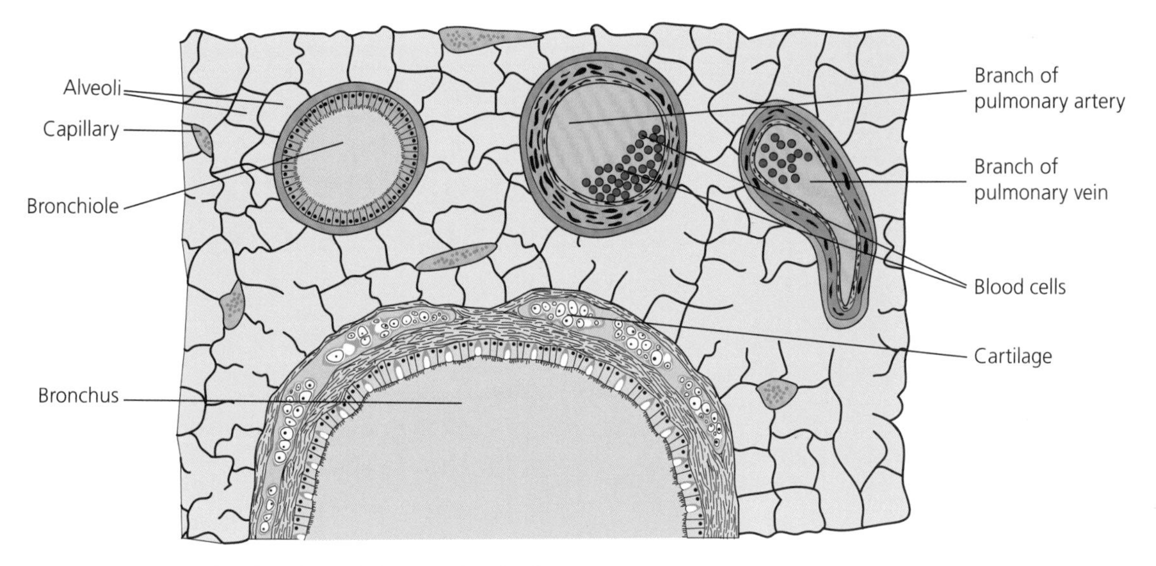

Figure 8.4 The distribution of tissues in the lungs

Ventilation

REVISED

Ventilation is also known as breathing. It refreshes the air in the alveoli. Ventilation is achieved by the action of the diaphragm and the intercostal muscles. The processes that take place during inspiration and expiration are summarised in Table 8.4.

Ventilation means breathing and refreshing the air in the alveoli.

Table 8.4 Inspiration and expiration

Structure/feature	Inspiration (inhaling)	Expiration (exhaling)
Diaphragm	Contracts and moves downwards, pushing organs down	Relaxes and is pushed up by the organs underneath
Intercostal muscles	Contract to raise the rib cage up and outwards	Relax and allow the rib cage to fall
Volume change	Chest cavity increases in volume	Chest cavity reduces in volume
Pressure change	Pressure inside chest cavity reduces and falls below atmospheric pressure	Pressure inside chest cavity rises above atmospheric pressure
Air movement	Air is pushed into lungs by higher atmospheric pressure	Air is pushed out of lungs by higher pressure in alveoli

Exam tip

To achieve full marks, you will need to describe all the volume and pressure changes accurately.

Revision activity

Draw a flow chart to describe inhaling and exhaling.

Vital capacity and tidal volume

REVISED

Vital capacity is the maximum volume of air that can be breathed in or out when taking a deep breath. Vital capacity is typically 4.5 dm^3 in young men and 3.0 dm^3 in young women. Vital capacity can be increased through training. Singers and athletes often have a large vital capacity.

Tidal volume is the volume of air breathed in and then out in one breath. The tidal volume changes according to the needs of the body. At rest it is usually about 0.5 dm^3.

Vital capacity is the maximum volume of air that can be breathed in or out in one breath.

Tidal volume is the volume of air breathed per breath, usually taken at rest.

Breathing rate and oxygen uptake

REVISED

Using a spirometer

A **spirometer** is an apparatus for measuring the volume of air inspired and expired by the lungs (Figure 8.5).

1 The subject should wear a nose clip to ensure that no oxygen escapes from the system and no additional air is added.
2 The subject breathes through the mouthpiece.
3 As the subject inhales, oxygen is drawn from the air chamber, which therefore descends.
4 As the subject exhales, the air chamber rises again.
5 Air returning to the air chamber passes through the canister of soda lime, which absorbs carbon dioxide.
6 The movements of the air chamber are recorded by a data logger or on a revolving drum.
7 Tidal volume is measured simply by allowing the subject to breathe normally.
8 Vital capacity is measured by asking the subject to breathe out as deeply as possible.

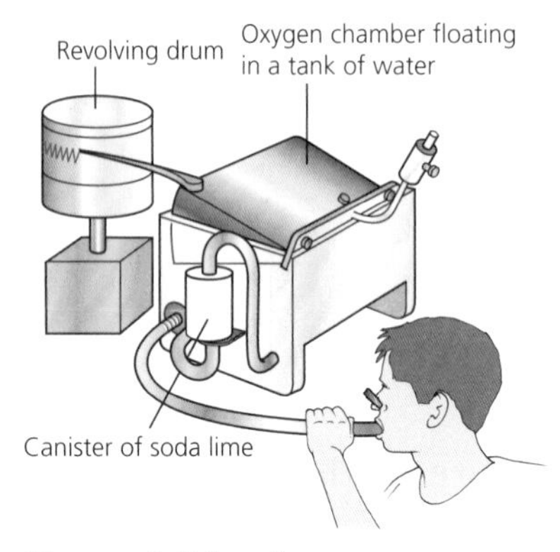

Figure 8.5 A spirometer

Analysing the trace

All measurements are taken from the spirometer trace (Figure 8.6). Always remember to measure at least three readings (if possible) and calculate a mean. **Breathing rate** is calculated by counting the number of peaks in 1 minute. **Oxygen uptake** is a little more difficult:

- As carbon dioxide is removed, the total volume in the air chamber decreases.
- The volume of oxygen absorbed is shown by the difference in height of the last peak from the first peak during normal breathing.
- Divide this volume by time taken in order to calculate the rate of oxygen uptake.

Now test yourself

4 Explain why the subject should wear a nose clip.
5 Explain the function of the soda lime and why it is essential.

Answers on p. 224

TESTED

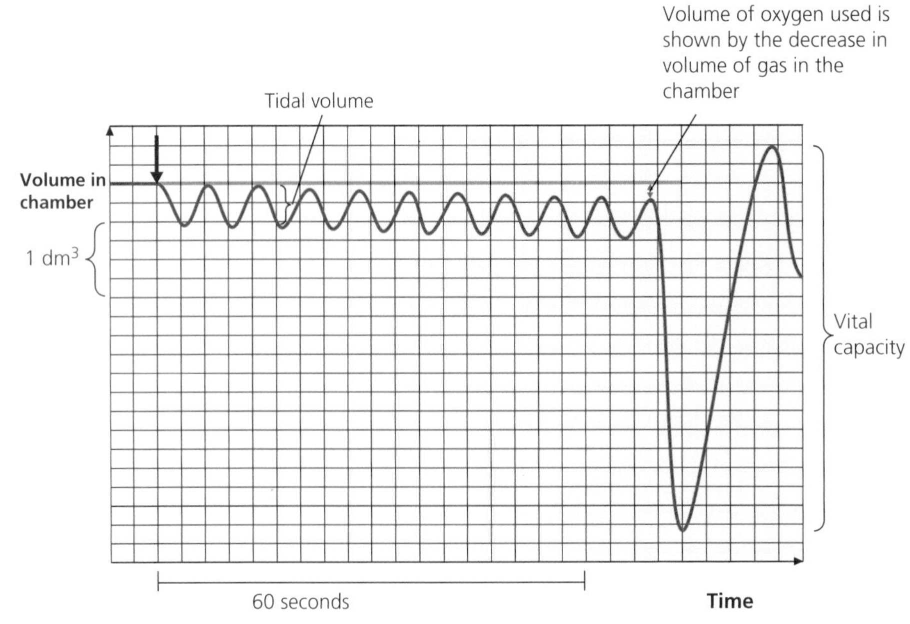

Figure 8.6 Measurements can be taken from a spirometer trace

Ventilation and gaseous exchange in bony fish and insects

Bony fish

REVISED

Fish must exchange gases with the water in which they live (Figure 8.7). They use gills to absorb oxygen dissolved in the water and release carbon dioxide into the water. Each gill consists of two rows of **gill filaments** (primary lamellae) attached to a bony arch. The filaments are very thin and their surface is folded into many **gill lamellae (gill plates)**. This provides a large surface area. Blood capillaries carry deoxygenated blood close to the surface of the gill plates where exchange takes place. The blood flows in the opposite direction to the flow of water (a **countercurrent flow**). Ventilation is achieved by movements of the floor of the mouth (**buccal cavity**) and **operculum** (the bony flap over the gills).

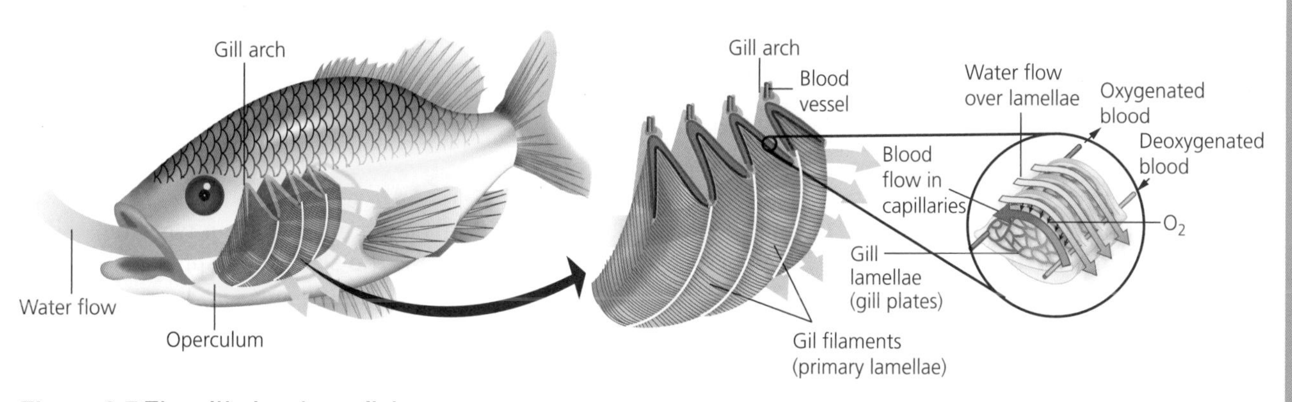

Figure 8.7 The gills in a bony fish

Insects

REVISED

Insects do not transport oxygen in blood. They have an air-filled tracheal system that supplies air directly to all the respiring tissues (Figure 8.8). Air enters the system via pores called **spiracles**. The air passes through the body in a series of tubes called tracheae (singular: **trachea**). These divide into smaller tubes called tracheoles. The ends of the tracheoles open into **tracheal fluid**. Gaseous exchange occurs between the air in the tracheole and the tracheal fluid.

Larger insects can ventilate their tracheal system by movements of the body which squeeze air sacs in the larger trachea.

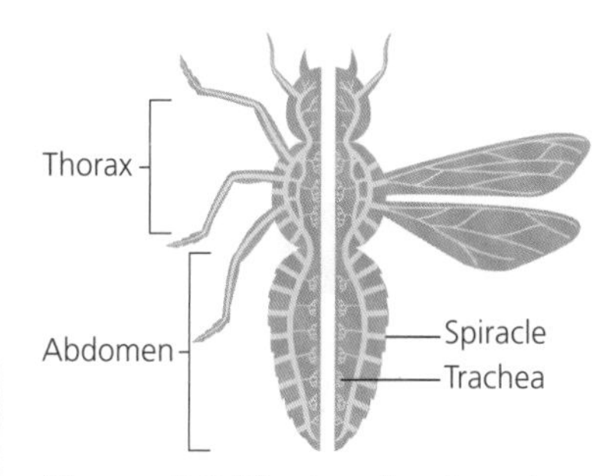

Figure 8.8 The tracheal system of an insect

Revision activity

Write a list of all the key terms used in this chapter, then add the meaning of each key term.

Exam practice

1 Which row in the following table correctly matches the function to the tissue? [1]

Row	Elastic tissue	Smooth muscle	Squamous epithelium	Ciliated epithelium
A	Reduce diameter of the lumen	Recoil to return alveoli to original size	Provide a short diffusion distance	Trap and remove bacteria
B	Recoil to return alveoli to original size	Reduce diameter of the lumen	Provide a short diffusion distance	Trap and remove bacteria
C	Contract to return alveoli to original size	Reduce diameter of the lumen	Provide a short diffusion distance	Trap and remove bacteria
D	Recoil to return alveoli to original size	Reduce diameter of the lumen	Trap and remove bacteria	Provide a short diffusion distance

2 (a) Explain why a large, active animal such as a mammal needs a specialised surface for gaseous exchange. [3]

(b) The following table shows how the surface area and volume of a sphere changes as its size increases.

Diameter (cm)	Surface area (cm^2)	Volume (cm^3)	Surface area to volume ratio
2	50.3	33.5	1.5 : 1
5	314.2	523.7	0.6 : 1
10	1256.8	4189.3	

(i) Calculate the surface area to volume ratio of a sphere of 10 cm diameter. Show your working. [2]

(ii) Describe the trend shown by the surface area to volume ratio as the size of the sphere increases. [2]

3 (a) State one function in the airways of each of the tissues listed below.
elastic tissue *ciliated epithelium* *smooth muscle* [3]

(b) Describe how ciliated cells and goblet cells work together to reduce the risk of infection in the lungs. [3]

(c) The alveoli walls contain elastic fibres. Suggest what may happen to the alveoli if the elastic fibres are damaged. [2]

(d) In asthmatics, certain substances in the air cause the smooth muscle in the walls of the airways to contract. Suggest the effect this will have on the person. [2]

4 (a) Describe how you would use a spirometer to measure tidal volume. [3]

(b) Explain why the air chamber should be filled with medical-grade oxygen rather than air. [2]

(c) Describe two other precautions that should be taken when using a spirometer. [2]

Answers and quick quiz 8 online

ONLINE

Summary

By the end of this chapter you should be able to:

- Understand the importance of surface area to volume ratios.
- Describe the features of a good gaseous exchange surface.
- Describe the features of the lungs that make them a good surface for gaseous exchange.
- Describe the distribution of tissues in the lungs and explain the role of each in an efficient organ of gaseous exchange.
- Outline the mechanism of breathing.
- Explain the terms *vital capacity* and *tidal volume*.
- Understand how a spirometer can be used to measure vital capacity, tidal volume, breathing rate and oxygen uptake.
- Describe the gaseous exchange system in bony fish and insects.

9 Transport in animals

The need for transport systems in multicellular animals

Size, metabolic rate and surface area to volume ratio

REVISED

All living animal cells need a supply of oxygen and nutrients to survive. They also need to remove waste products such as carbon dioxide and urea so that they do not build up and become toxic. The three main factors that affect the need for a transport system are **size**, **metabolic rate** and **surface area to volume ratio**. You should recall that size and surface area to volume ratio have been explained in Chapter 8. Active organisms usually have a high metabolic rate. This requires more oxygen to allow more aerobic respiration to take place. This is essential so that more ATP can be released from food to provide energy for the higher level of activity. The oxygen and substrate molecules such as glucose must be supplied to the active cells rapidly, which is why active organisms require a transport system.

Typical mistake

Many candidates describe small organisms as having a large surface area rather than a large surface area to volume ratio, or they describe a large organism as having a small surface area.

Exam tip

Don't confuse surface area with surface area to volume ratio. It may help to always write SA : V rather than surface area to volume ratio.

Different types of circulatory systems

Single and double circulatory systems

REVISED

In a **single circulatory system**, blood flows through the heart once every time it goes around the body. Fish have a single circulatory system. The blood flows from the heart to the gills and then on to the body before returning to the heart:

heart → gills → body → heart

In a **double circulatory system**, blood flows through the heart twice for every circuit around the body. Mammals have developed a circulation that involves two separate circuits. One circuit carries blood to the lungs to take up oxygen. This is the pulmonary circulation. The other circuit carries the oxygen and nutrients around the body to the tissues. This is the systemic circulation. The heart is adapted to form two pumps, one for each circulation.

body → heart → lungs → heart → body

Typical mistake

Many candidates describe a single circulation as 'blood going through the heart once'. However, the blood must go around the body and return to the heart, so it is better to describe it as 'going through the heart once for every circuit of the body'.

Now test yourself

TESTED

1 Explain why a double circulatory system is more efficient than a single circulatory system.

Answer on p. 224

Open and closed circulatory systems

REVISED

Insects have an **open circulatory system**. In an open system:

1 there is no separate tissue fluid
2 blood circulates around the organs and cells
3 pressure cannot be raised to help circulation
4 circulation is affected by body movements
5 oxygenated and deoxygenated blood mix freely

Fish and **mammals** both have a **closed circulatory system**. In a closed system:

- blood is kept in vessels
- pressure can be maintained
- pressure can be higher
- flow can be faster
- flow can be directed to certain tissues or organs

Revision activity

Draw an insect and a mammal. Inside each illustration, include a simple version of the circulation system.

The structure and functions of blood vessels

Arteries, arterioles, capillaries, venules and veins

REVISED

Blood flows through a series of vessels. Each is adapted to its particular role in relation to its distance from the heart. All types of blood vessel have an inner layer or lining made of cells called the **endothelium**. This is a thin layer that is particularly smooth to reduce friction with the flowing blood. Figure 9.1 shows cross-sections of an artery and a vein. Table 9.1 compares the structure and functions of **arteries**, **veins** and **capillaries**.

The **endothelium** is a thin layer of cells that lines all blood vessels.

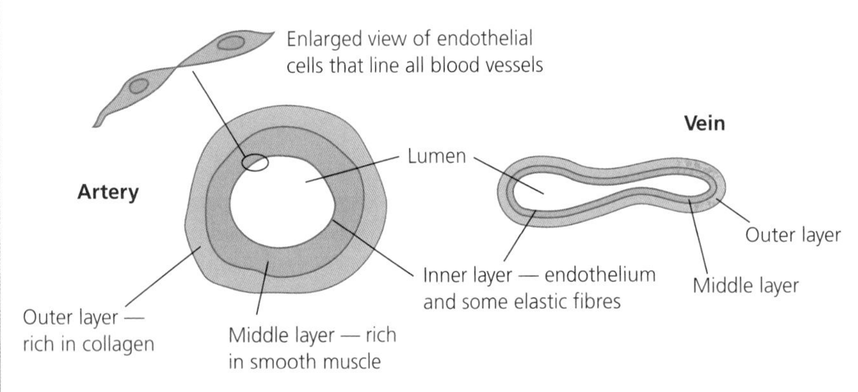

Figure 9.1 Cross-sections of an artery and a vein

Arterioles are small arteries with a spiral layer of smooth muscle. They distribute blood from the arteries to the capillaries and can be constricted to reduce blood flow. **Venules** are small veins that collect the blood from the capillaries and lead into the veins.

Table 9.1 Comparing the structure and functions of arteries, veins and capillaries

Feature	Arteries	Veins	Capillaries
Function	Transport the blood away from the heart	Transport the blood back to the heart	Enable the exchange of materials between the blood and tissue fluid
Thickness of wall	Thick	Thin	Very thin (one cell thick)
Components of wall	Endothelium lining surrounded by a thick middle layer of elastic tissue and smooth muscle, then a thick outer layer rich in collagen	Endothelium lining surrounded by a thinner layer of elastic tissue and smooth muscle, with a thin outer layer containing collagen	One layer of endothelium cells
Blood pressure	High	Low	Low
Presence of valves	No	Yes	No
Cause of flow	Pressure created by the heart, maintained by recoil of elastic tissues	Squeezing action of body muscles and valves to ensure correct direction	Pressure from action of the heart

Typical mistake

Many candidates describe the role of smooth muscle as 'creating a smooth surface to reduce friction', or they describe the action of the smooth muscle as 'pumping blood along the artery'. The smooth muscle is used to reduce the diameter of the vessel to maintain blood pressure or to reduce blood flow in a particular direction. Similarly, the elastic tissue is used to recoil the artery wall to maintain blood pressure.

Revision activity

Write a list the features of the arteries that maintain blood pressure and another list of the features that enable arteries to withstand high blood pressure.

Now test yourself

TESTED

2 Describe how the structure of an artery is adapted to its role of transporting blood at high pressure.
3 Describe how the structure of a vein is adapted to its role of transporting blood at low pressure.

Answers on p. 224

Blood, tissue fluid and lymph

Blood

REVISED

Blood is the fluid found inside the blood vessels. It consists of:

- water-based plasma containing dissolved substances — oxygen, nutrients such as glucose and amino acids, lipoproteins, carbon dioxide (most transported as bicarbonate ions), other wastes such as urea, hormones, antibodies and plasma proteins
- red blood cells (erythrocytes), probably carrying oxygen
- white blood cells (phagocytes), such as neutrophils and lymphocytes
- platelets

Tissue fluid

REVISED

Tissue fluid surrounds the body cells. It is the plasma that has been filtered out of the blood, so it contains all the dissolved elements of the blood except the cells, platelets and plasma proteins. These are too large to pass out of the blood vessels. There may be some phagocytic neutrophils in tissue fluid as these can change shape to squeeze out of the blood vessels.

Exam tip

Remember that tissue fluid is blood that does not contain blood cells or plasma proteins, but it does contain the dissolved components.

The formation of tissue fluid

The walls of the capillaries are a single layer of endothelium cells. Fluid and dissolved substances in the fluid can squeeze between the endothelium cells. The fluid is acted upon by two forces:

1. the **hydrostatic pressure** gradient between the blood and the tissue fluid, which tends to push fluid out of the capillary
2. the **oncotic pressure** gradient between the blood and tissue fluid, which tends to move fluid into the blood because the water potential of the blood is lower than water potential of the tissue fluid

Hydrostatic pressure is the pressure a fluid exerts on the sides of a vessel.

Oncotic pressure is the osmotic pressure created by dissolved substances such as proteins.

At the arterial end of the capillary, the hydrostatic pressure created by the heart is still quite high. Therefore, there is a steep pressure gradient pushing fluid out of the capillary. This overcomes the oncotic pressure gradient and fluid moves out of the capillary to become tissue fluid. Oxygen and nutrients move into the tissues with the fluid.

At the venous end of the capillary, the hydrostatic pressure is lower. The hydrostatic pressure gradient is less steep than the oncotic pressure gradient and fluid returns to the capillary. Carbon dioxide and other wastes are carried back into the blood as tissue fluid moves back into the capillary. Some of the tissue fluid is drained into the blind-ending lymph vessels to become lymph.

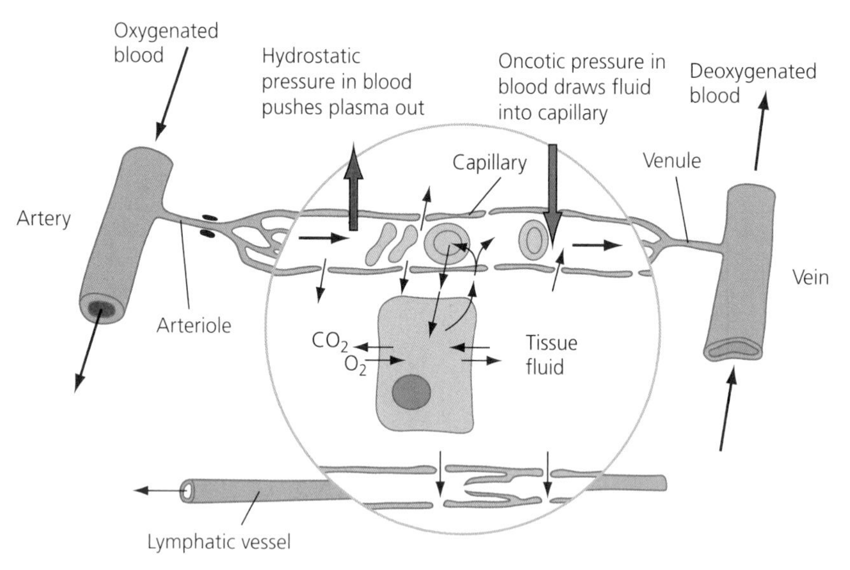

Figure 9.2 The flow of blood through a capillary

Lymph

REVISED

Lymph is excess tissue fluid that is not returned to the blood vessel. Instead, it is drained into the lymph vessels. These carry the fluid back to the circulatory system by a different route.

Lymph contains the same substances as tissue fluid but it has less oxygen and glucose as these have been used by the cells. Lymphocytes produced in the lymph nodes may also be present.

Revision activity

Draw a table to compare the compositions of blood, tissue fluid and lymph.

The mammalian heart

The **mammalian heart** is a muscular pump that is divided into two sides. The right side pumps **deoxygenated blood** to the lungs to be oxygenated. The left side pumps **oxygenated blood** to the rest of the body. On both sides the action of the heart is to squeeze the blood, putting it under pressure and forcing it along the arteries.

Deoxygenated blood transports less oxygen and more carbon dioxide than oxygenated blood.

Oxygenated blood transports oxygen to the organs.

External structure

REVISED

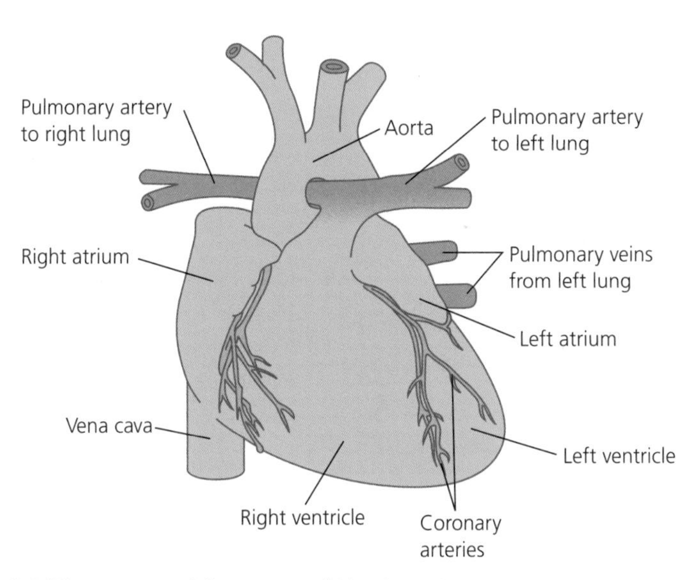

Figure 9.3 The external features of the heart

The muscle surrounding the two main pumping chambers (the ventricles) is dark red. Above the ventricles are two thin-walled chambers known as the atria (singular: atrium). These are much smaller than the ventricles. On the surface of the heart are coronary arteries. These carry oxygenated blood to the heart muscle itself. These arteries are important as the heart continually works hard. If they become constricted, blood flow to the heart muscle is restricted and this can reduce the delivery of oxygen and nutrients, such as fatty acids and glucose, causing a heart attack. At the top of the heart are the veins that carry blood into the heart and the arteries that carry blood out of the heart.

Revision activity

Draw a diagram to show the external features of the heart and annotate it with the names and functions of each feature.

Internal structure

REVISED

The heart is divided into four chambers: two atria and two ventricles.

The atria

The two upper chambers are **atria**. These receive blood from the major veins. Deoxygenated blood flows from the vena cava into the right atrium. Oxygenated blood flows from the pulmonary vein into the left atrium. The atria have very thin walls as they do not need to create much pressure. Blood simply flows through the atria into the ventricles. When the ventricles are nearly full, the atrial walls contract just to completely fill the ventricles.

The **atria** are the small chambers at the top of the heart, which collect blood from the main veins.

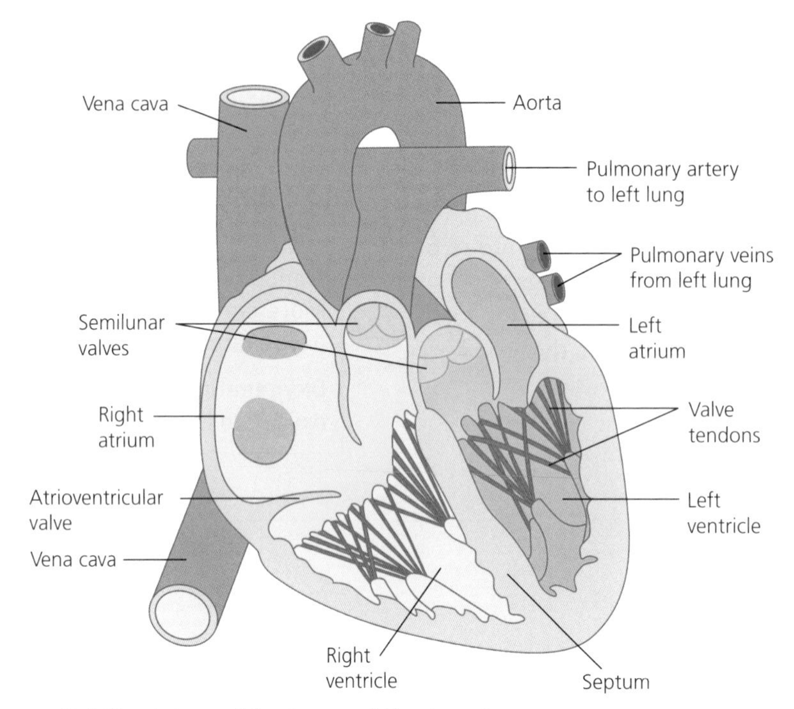

Figure 9.4 The internal features of the heart

The ventricles

The two lower chambers are the **ventricles**. Each has a thick muscular wall. The wall contracts to create pressure which pushes the blood into the arteries.

The right ventricle has walls that are thicker than the atrial walls. This enables it to pump blood out of the heart. The right ventricle pumps deoxygenated blood to the lungs. Therefore, the blood does not need to travel far. Also, the lungs contain many fine capillaries. The pressure of the blood must not be too high to prevent the capillaries in the lungs bursting.

The walls of the left ventricle are often two or three times thicker than those of the right ventricle. The blood from the left ventricle is pumped out through the aorta and needs sufficient pressure to propel it all the way to the extremities of the body. The pressure created by the left ventricle is typically measured at 110–120 mmHg.

The ventricles are separated from each other by a wall of muscle called the **septum**. This ensures that the oxygenated blood in the left side of the heart and deoxygenated blood in the right side are kept separate.

The **ventricles** are the larger lower chambers, which have thick walls to pump the blood out of the heart.

The **septum** is the muscle that separates the two ventricles.

Revision activity

Draw a diagram to show the internal features of the heart and annotate it with the names and functions of each feature.

Now test yourself

TESTED

4 Explain why the left ventricle looks so much bigger than the right.

Answer on p. 224

Major arteries

Deoxygenated blood leaving the right ventricle flows into the pulmonary artery leading to the lungs. Oxygenated blood leaving the left ventricle flows into the aorta. This carries blood to a number of arteries that supply all parts of the body.

The cardiac cycle

REVISED

It is important that the chambers of the heart all contract in a coordinated fashion. If the chambers were to contract out of sequence, this would lead to inefficient pumping. The sequence of events involved in one heart beat is called the **cardiac cycle** (Figure 9.5).

The **cardiac cycle** is the series of events in one heart beat.

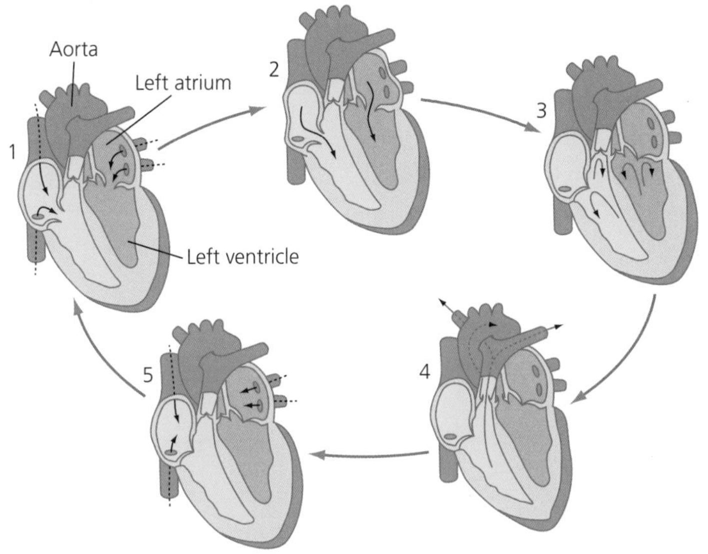

Aorta
Left atrium
Left ventricle
1
2
3
4
5

Figure 9.5 The cardiac cycle

1 Blood returns to the heart from the body (via the vena cava) and lungs (via the pulmonary vein). Both the atria fill at the same time. The **valves** between the atria and the ventricles (the **atrioventricular valves**) are open to allow blood to flow straight through into the ventricles.
2 Once the ventricles are nearly full, the **sinoatrial node (SAN)** initiates a new heartbeat. It creates a wave of excitation which spreads over the walls of the atria. The walls contract, pushing a little extra blood from the atria into the ventricles. The wave of excitation is stopped by a layer of non-conducting fibres between the atria and the ventricles. The wave of excitation can only pass through the **atrioventricular node (AVN)**, where it is delayed a little. This allows time for the ventricles to fill.
3 After the delay, the wave of excitation passes down the bundle of His in the septum between the ventricles. At the base of the septum, the bundle splits into separate fibres called **Purkyne tissue** that carry the excitation up the walls of the ventricles, causing contraction from the base (apex) upwards. The walls of the two ventricles contract together. As the pressure rises, the atrioventricular valves are pushed shut which prevents blood re-entering the atria. The tendinous cords attached to the valves prevent them from inverting.
4 The blood pressure in the ventricles rises quickly until it rises above the pressure in the aorta and pulmonary artery. This pushes the **semilunar valves** open and blood is pushed into the main arteries.
5 Once contraction is complete, the muscles relax and the elasticity of the walls causes recoil to return the ventricles to their original size and shape. This causes the pressure in the ventricles to drop quickly. When the pressure drops below the pressure in the main arteries, the semilunar valves are pushed shut, preventing re-entry of blood into the ventricles. When the pressure in the ventricles drops below the pressure in the atria, the atrioventricular valves are pushed open by the pressure in the atria. This allows blood to flow into the ventricles again.

The **atrioventricular valves** lie between the atria and the ventricles.

The **semilunar valves** lie at the entrance to the main arteries.

Revision activity

Draw an outline of the heart and add arrows to show the direction of blood flow.

Now test yourself

TESTED

5 Explain how the atrioventricular valves are opened and closed.
6 Explain why the electrical stimulation wave must be delayed at the AVN.

Answers on p. 224

Pressure changes during contraction

The changes in pressure in the heart can be represented by a graph (Figure 9.6). The important points are where one line crosses another — this is where the pressure in one chamber rises above that in another chamber, causing a valve to open or close.

Exam tip

Remember that it is the walls of the atria and ventricles that contract.

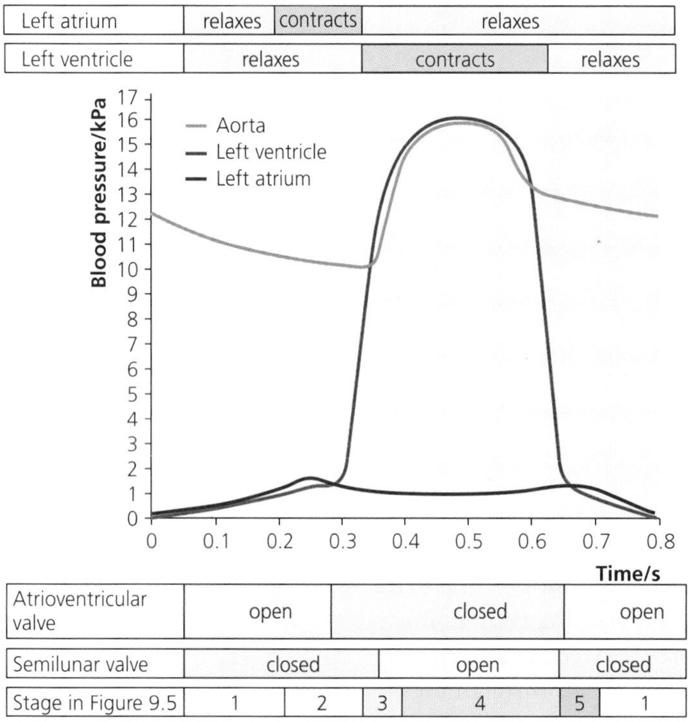

Left atrium	relaxes	contracts	relaxes
Left ventricle	relaxes	contracts	relaxes

Atrioventricular valve	open	closed	open
Semilunar valve	closed	open	closed

Stage in Figure 9.5	1	2	3	4	5	1

Figure 9.6 The pressure changes in the heart

Now test yourself

TESTED

7 Describe and explain what happens at the point in Figure 9.6 where the line for the pressure in the left ventricle rises above the line for the pressure in the left atrium.

Answer on p. 224

Electrocardiograms

REVISED

An **electrocardiogram (ECG)** (Figure 9.7) records the electrical activity of the heart. Wave P is the excitation of the atria. Wave QRS is the excitation of the ventricles. Wave T is associated with ensuring the muscles have time to rest. Abnormal heart activity can often be identified by an abnormal ECG trace. The waves may be smaller, inverted or further apart.

Exam tip

Questions are likely to show you a normal trace and an abnormal trace and ask you identify the differences.

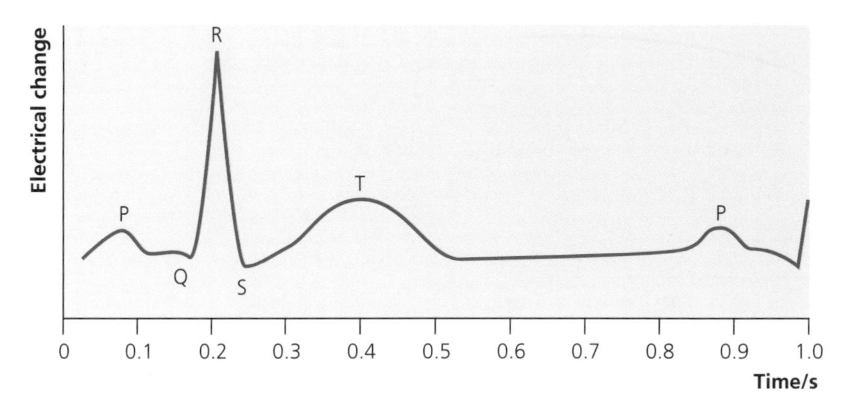

Figure 9.7 A normal ECG trace

The role of haemoglobin in transporting oxygen and carbon dioxide

Transport of oxygen

REVISED

Oxygen enters the blood in the lungs. Oxygen molecules diffuse into the blood plasma and red blood cells. Here, they associate with the **haemoglobin** (Hb) to form **oxyhaemoglobin**.

Haemoglobin is a complex protein with four subunits. Each subunit contains a haem group that contains a single iron ion (Fe^{2+}). This attracts and holds one oxygen molecule. The haem group is said to have an **affinity** for (an attraction to) oxygen — the haemoglobin attracts and holds the oxygen. Each haem group can hold one oxygen molecule, so each haemoglobin molecule can carry four oxygen molecules.

Haemoglobin is the red pigment that transports oxygen.

Oxyhaemoglobin is the product that is formed when oxygen from the lungs combines with haemoglobin in the blood.

Oxygen dissociation curves

Haemoglobin has a high affinity for oxygen. The amount it takes up depends on the amount of oxygen in the surrounding tissues, which is measured by its partial pressure of oxygen (pO_2) or oxygen tension.

At a low pO_2, haemoglobin does not readily take up oxygen molecules. The haem groups are hidden at the centre of the haemoglobin molecule, making it difficult for the oxygen molecules to associate with them. This difficulty in combining the first oxygen molecule accounts for the low saturation level of haemoglobin at low pO_2.

As the pO_2 rises, one oxygen molecule succeeds in associating with one of the haem groups. This causes a conformational change in the shape of the haemoglobin molecule, which allows more oxygen molecules to associate with the other three haem groups more easily. This accounts for the steepness of the curve as pO_2 rises (Figure 9.8). In the body tissues, cells need oxygen for aerobic respiration, so oxyhaemoglobin releases the oxygen. This is called **dissociation** and it is shown in an **oxygen dissociation curve**.

Dissociation is the release of oxygen from oxyhaemoglobin.

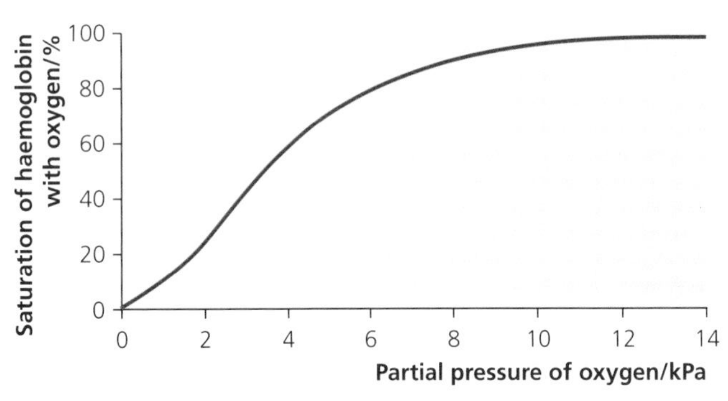

Figure 9.8 An adult oxygen dissociation curve

Now test yourself

TESTED

8 Explain why a person using a spirometer containing air would soon feel tired and breathless.

Answer on p. 225

Fetal haemoglobin

The haemoglobin of a mammalian fetus has a higher affinity for oxygen than adult haemoglobin. Its haemoglobin must be able to take up oxygen from an environment that makes the adult haemoglobin release oxygen. In the placenta, the **fetal haemoglobin** must absorb oxygen. This reduces the oxygen tension near the blood, making the maternal haemoglobin release oxygen. Therefore, the oxyhaemoglobin dissociation curve for fetal haemoglobin is to the left of the curve for adult haemoglobin.

Fetal haemoglobin is a modified form of haemoglobin found in the mammalian fetus.

Transport of carbon dioxide

REVISED

Carbon dioxide released from respiring tissues must be removed from the tissues and transported to the lungs. Carbon dioxide in the blood is transported in three ways:

- as hydrogencarbonate ions (HCO_3^-) in the plasma (85%)
- combined directly with haemoglobin to form a compound called carbaminohaemoglobin (10%)
- dissolved directly in the plasma (5%)

How are hydrogencarbonate ions formed?

As carbon dioxide diffuses into the blood, some of it diffuses into the red blood cells. Here, it combines with water to form a weak acid (carbonic acid). This reaction is catalysed by the enzyme **carbonic anhydrase**.

$$CO_{2+} H_2O \rightarrow H_2CO_3$$

The carbonic acid dissociates to release hydrogen ions (H^+) and hydrogencarbonate ions (HCO_3^-).

$$H_2CO_3 \rightarrow H^+ + HCO_3^-$$

The hydrogencarbonate ions diffuse out of the red blood cells into the plasma. The charge inside the red blood cells is maintained by the movement of chloride ions (Cl^-) from the plasma into the red blood cells. This is called the **chloride shift**.

The hydrogen ions could cause the contents of the red blood cells to become acidic. To prevent this, the hydrogen ions are taken up by haemoglobin to produce **haemoglobinic acid**. The haemoglobin is acting as a buffer (a compound that can maintain a constant pH).

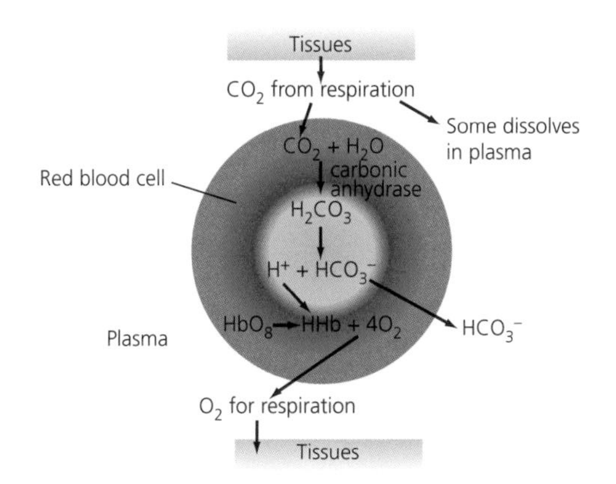

Figure 9.9 How carbon dioxide is converted to hydrogencarbonate ions

> **Exam tip**
>
> Ensure you get all the details correct, including the names and the chemical formulae.

The Bohr effect

The hydrogen ions produced are absorbed by the haemoglobin. However, in very active tissues a lot of carbon dioxide is released and, therefore, a lot of hydrogen ions are created. As the concentration of hydrogen ions increases, this decreases the pH of the cytoplasm. The change in pH alters the structure of the haemoglobin, reducing its affinity for oxygen. Therefore, more oxygen is released. The haemoglobin dissociation curve shifts to the right (**the Bohr effect**).This also frees up more haemoglobin molecules to absorb the additional hydrogen ions.

> **The Bohr effect** is the shift to the right of the position of the haemoglobin dissociation curve in the presence of extra carbon dioxide.

> **Revision activity**
>
> Write a list of all the key terms used in this chapter, then add the meaning of each key term.

Exam practice

1 (a) Describe the initiation and coordination of one heart beat. [5]

(b) (i) In a condition known as superventricular tachycardia, the non-conducting fibres between the atria and the ventricles occasionally conduct the excitation wave. Suggest what effect this might have on the action of the heart. [2]

(ii) Suggest what effects this might have on the patient. [2]

2 Explain why artery walls contain both smooth muscle and elastic fibres. [4]

3 (a) State what creates the hydrostatic pressure at the arterial end of the capillary. [2]

(b) State two reasons why the pressure is much lower at the venous end of the capillary. [2]

4 Describe what is meant by an open circulatory system and explain why it is less efficient than a closed system. [3]

Answers and quick quiz 9 online

ONLINE

Summary

By the end of this chapter you should be able to:

- Explain the need for transport systems in large and active organisms.
- Explain the difference between open/closed circulatory systems and single/double circulatory systems.
- Describe the structure of the blood vessels.
- Understand the differences between blood, tissue fluid and lymph.
- Describe the external and internal structures of the mammalian heart.
- Describe the action of the heart and the cardiac cycle.
- Understand how oxygen and carbon dioxide are transported.

10 Transport in plants

The need for transport systems in multicellular plants

Size, metabolic rate and surface area to volume ratio

REVISED

The need for a transport system in large and active organisms is explained in Chapter 9.

The structure and function of the vascular system

Roots, stem and leaves

REVISED

Multicellular plants are large organisms so they need a transport system (Figure 10.1). Xylem tissue moves water and minerals from the roots to the leaves. Phloem tissue moves assimilates up and down the plant from sources to sinks.

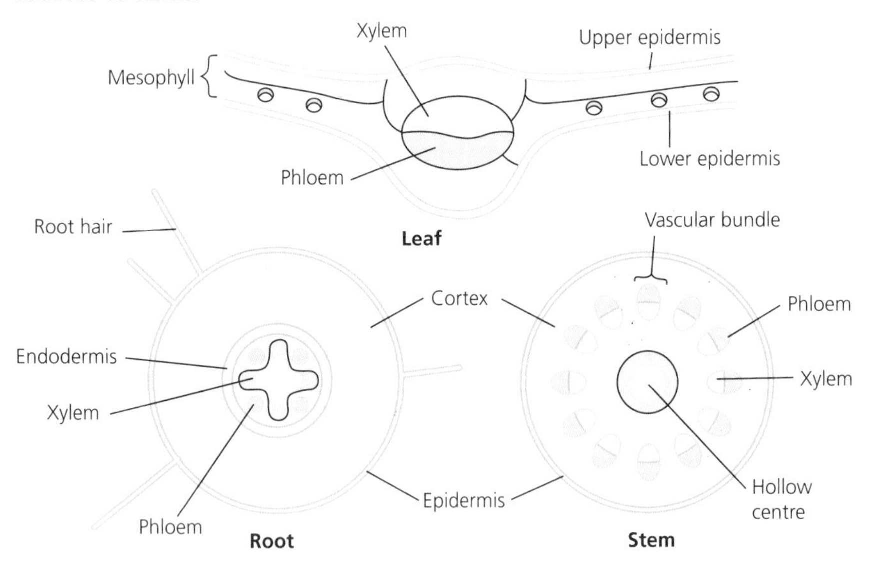

Figure 10.1 The distribution of transport tissues in a leaf, stem and root

Revision activity

From memory, draw the position of the xylem and phloem in roots and stems.

Xylem vessels

REVISED

Xylem vessels are adapted to enable the free flow of water along the vessels. The cells have been killed by impregnation of the walls with lignin.

Adaptations of xylem

Xylem vessels have several adaptations (Figure 10.2):

- end walls are removed to form long tubes

- no cytoplasm or organelles are present
- cell walls are impregnated with lignin (lignified) to make the vessel wall waterproof and to strengthen the vessel to prevent it collapsing
- spiral, annular and reticulate thickening strengthens the wall to prevent collapse
- bordered pits between the vessels allow the movement of water between vessels

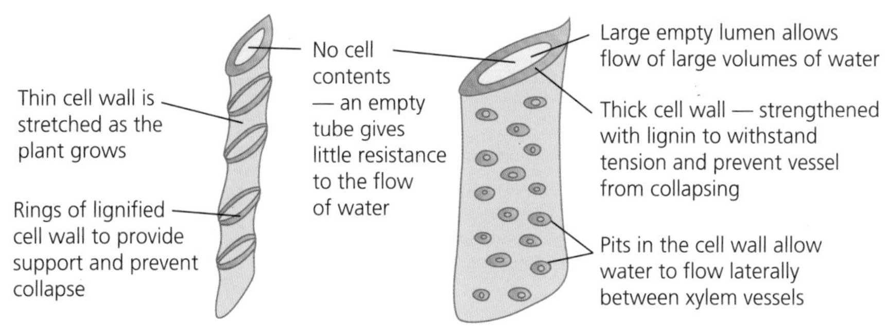

Thin cell wall is stretched as the plant grows
Rings of lignified cell wall to provide support and prevent collapse
No cell contents — an empty tube gives little resistance to the flow of water
Large empty lumen allows flow of large volumes of water
Thick cell wall — strengthened with lignin to withstand tension and prevent vessel from collapsing
Pits in the cell wall allow water to flow laterally between xylem vessels

Figure 10.2 Xylem vessels showing adaptations

Exam tip

Make sure you say that the *cell walls* have been lignified — don't say that the vessels or the xylem have been lignified. Remember that this makes the walls waterproof — the xylem itself is not waterproof as there are pits to allow water in and out.

Revision activity

Draw a diagram of a xylem vessel and annotate it with the features that adapt it to transporting water.

Now test yourself

TESTED

1 Explain why the xylem vessel walls are impregnated with lignin and why it is important to have pits in the walls.

Answer on p. 225

Sieve tube elements and companion cells

REVISED

The phloem is adapted to transport assimilates actively by mass flow. There are two cell types involved:

- **sieve tube elements** — long sieve tubes that transport the assimilates
- **companion cells** — support cells that provide all the metabolic functions for the sieve tube elements and are involved in actively loading the sieve tubes

Adaptations of phloem

Phloem cells have several adaptations, as summarised in Table 10.1.

Table 10.1 Adaptations of phloem cells

Cell	Adaptations
Sieve tube elements	Form long tubes End walls are retained End walls contain many sieve pores, so they are called sieve plates Thin layer of cytoplasm Very few organelles and no nucleus
Companion cells	Closely associated with sieve tube elements Connected to sieve tube elements by many plasmodesmata Dense cytoplasm with many mitochondria Large nucleus

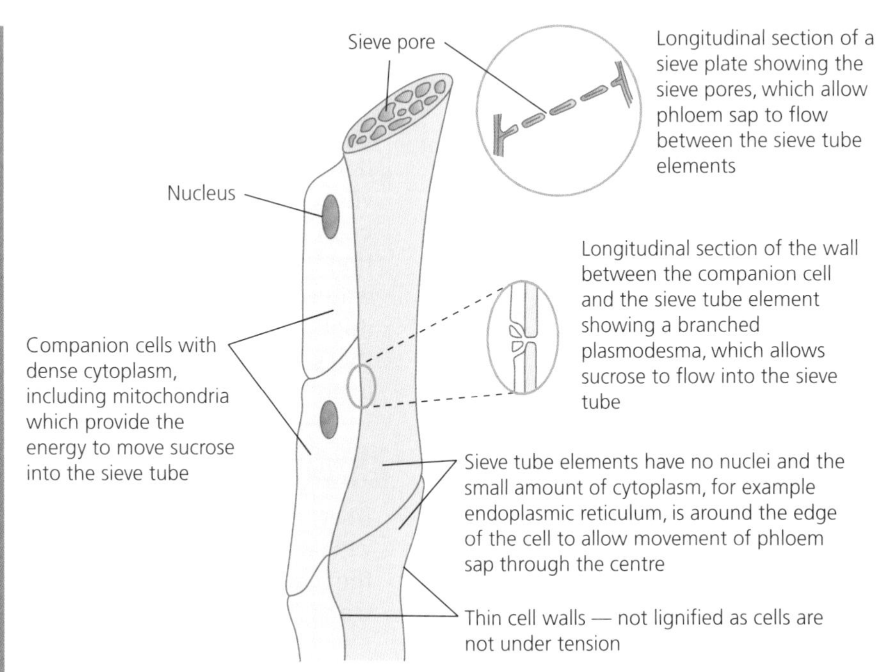
Sieve pore
Longitudinal section of a sieve plate showing the sieve pores, which allow phloem sap to flow between the sieve tube elements
Nucleus
Longitudinal section of the wall between the companion cell and the sieve tube element showing a branched plasmodesma, which allows sucrose to flow into the sieve tube
Companion cells with dense cytoplasm, including mitochondria which provide the energy to move sucrose into the sieve tube
Sieve tube elements have no nuclei and the small amount of cytoplasm, for example endoplasmic reticulum, is around the edge of the cell to allow movement of phloem sap through the centre
Thin cell walls — not lignified as cells are not under tension

Figure 10.3 Phloem sieve tubes and companion cells showing adaptations

Revision activity

Draw a length of phloem tissue and annotate it with the features that adapt it to transporting assimilates.

Now test yourself

2 Explain why companion cells are essential.

Answer on p. 225

TESTED

The process of transpiration

Transpiration as a consequence of gaseous exchange

REVISED

Transpiration is the loss of water vapour from the upper parts of a plant, mainly the leaves. Although some water evaporates and diffuses through the leaf surface, most water vapour is lost via the stomata. Transpiration involves three stages:

1 Water moves by osmosis from the xylem to the mesophyll cells in the leaf.
2 Water evaporates from the surfaces of the spongy mesophyll cells into the air spaces inside the leaves.
3 Water vapour diffuses out of the leaf via the stomata.

The stomata open during the day to allow gaseous exchange — carbon dioxide enters the leaf and oxygen is released. This is to enable photosynthesis to occur. As the stomata are open, water vapour is lost. Transpiration is therefore a consequence of gaseous exchange.

Transpiration is the loss of water vapour from the aerial parts of the plant.

Typical mistake

Many candidates state that water is lost from the leaf — it is *water vapour* that is lost.

Factors that affect the rate of transpiration

There must be a water potential gradient between the air spaces in the leaf and the surrounding air to make water vapour leave the leaf. The steeper this gradient, the more rapid the loss of water vapour (transpiration). Factors that increase transpiration rate include:

- higher temperatures — this increases evaporation so there will be a higher water potential inside the leaf
- more wind — this blows water vapour away from the leaf, reducing the water potential in the surrounding air
- lower relative humidity — this increases the water potential gradient between the air inside the leaf and outside
- higher light intensity — this causes the stomata to open wider

Revision activity

Draw a cross-section of a leaf and add arrows to show the movement of water molecules. Label each arrow and explain what is happening to make the water to move.

Exam tip

You should explain transpiration using the term *water potential gradient*.

Now test yourself

TESTED

3 Explain why transpiration is quicker on a hot sunny day than on a cool cloudy day.

Answer on p. 225

Typical mistake

Many candidates confuse water potential and water potential gradient. A gradient must be between two points such as between the air inside the leaf and the air outside.

Measuring the rate of transpiration

Transpiration can be estimated using a bubble **potometer** (Figure 10.4). A potometer actually measures water uptake by the stem, but you can assume that water uptake equals water loss from the leaves in most cases. Care must be taken when setting up the potometer to ensure that there are no leaks and no air in the system, except the bubble used for measuring. Once the shoot has been allowed to acclimatise, the movement of the bubble along the capillary can be measured under different conditions. Transpiration rate is calculated by dividing the distance moved by a set time.

Exam tip

To gain the marks, you must stress that higher temperatures increase transpiration, not simply that temperature increases transpiration.

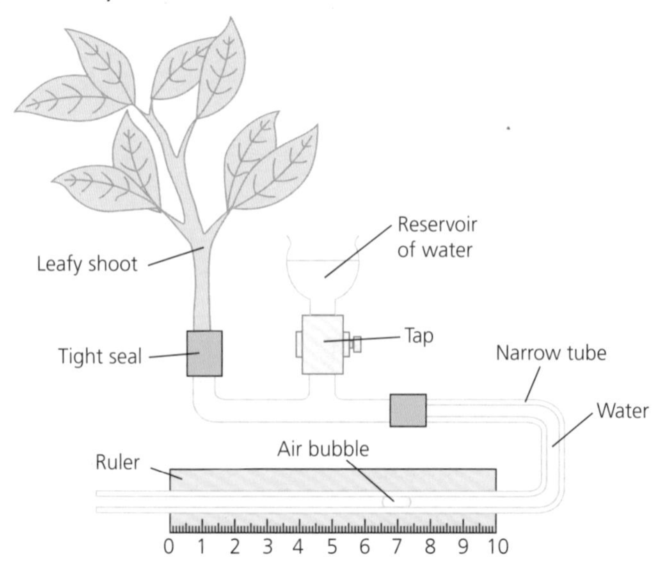

Figure 10.4 A bubble potometer

Revision activity

List the precautions you should take when setting up a bubble potometer.

The transport of water

Transport of water

REVISED

The cell wall of a plant cell is permeable to water. The cell-surface membrane is selectively permeable. As a result, plant cells can contain mineral ions in solution which reduces the **water potential** inside the cell. The more concentrated the mineral ions, the lower the water potential. Water moves by osmosis from a cell with a higher water potential to a cell with a lower water potential because water molecules move down their water potential gradient, as shown in Figure 10.5. The arrows show the direction of water movement between three mesophyll cells in a leaf. Cell P has the highest water potential and the water potential of cell Q is higher than that of cell R.

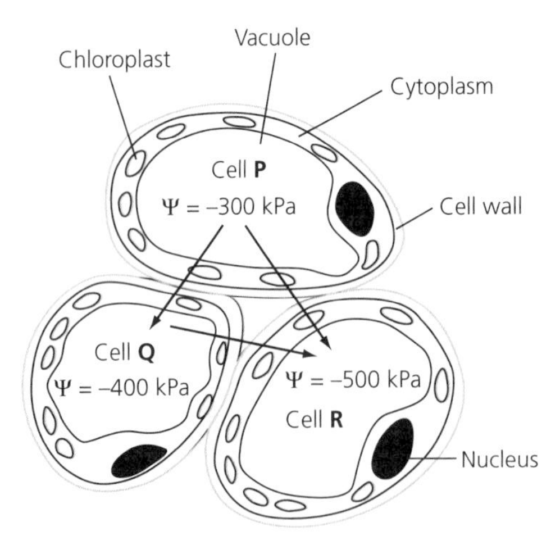

Figure 10.5 Water moves from cell to cell down a water potential gradient

Similarly, water can enter a cell from its environment if the water potential in the cell is lower than the water potential in the environment. This is how root hair cells absorb water from the soil.

Typical mistake

Many students treat –1200 kPa as higher than –1100 kPa. Remember that the figures are negative. It may be best to describe one as more negative than the other.

Pathways

Once in the plant, water can move across a tissue such as the root cortex by different **pathways** (Figure 10.6):

1. the apoplast pathway carries water between the cells through the cell walls — the water does not enter the cytoplasm or pass through cell-surface membranes
2. the symplast pathway takes water from cell to cell through the cytoplasm of each cell. Water often passes through plasmodesmata linking the cytoplasm of adjacent cells
3. the vacuolar pathway carries water through the cytoplasm and vacuole of each cell

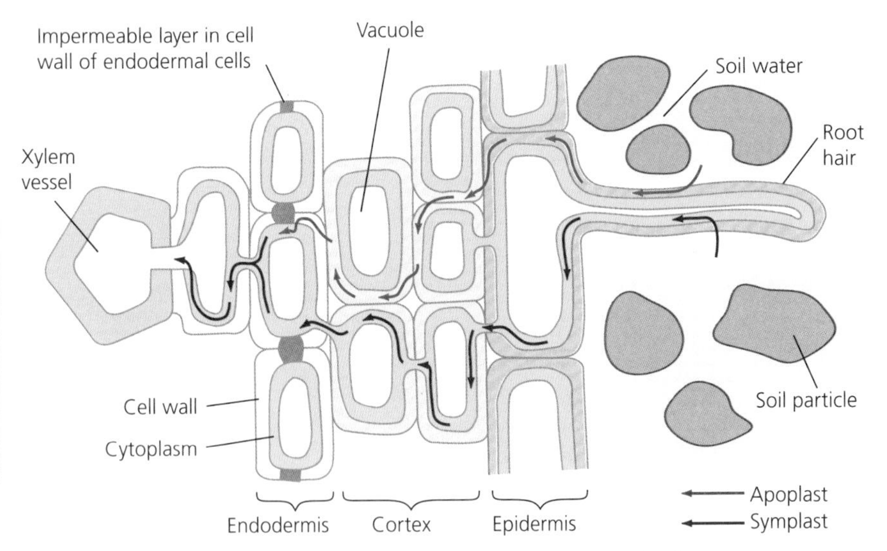

Figure 10.6 Water enters the root hair cells by osmosis. It may pass across the root by the apoplast pathway or the symplast pathway

Exam tip

Remember that transpiration is the loss of water vapour from the leaf. The transpiration stream is the flow of water from the roots to the leaves to replace the water lost in transpiration.

The transpiration stream

Water movement from the roots up to the leaves in the xylem is known as the **transpiration stream**. There are three mechanisms that move water up the stem: root pressure, adhesion or capillary action and transpirational pull. Root pressure and capillary action combined can only raise water by a few metres. Therefore, transpiration and the pull it creates are essential to move water all the way up a tall stem.

The **transpiration stream** is the movement of water from the roots to the leaves.

Now test yourself

TESTED

4 Explain the difference between transpiration and the transpiration stream.

Answer on p. 225

Root pressure

Root pressure is created by the action of the endodermis in the roots. The endodermis uses metabolic energy to pump mineral ions into the root medulla. This reduces the water potential in the medulla and xylem, making it more negative than in the cortex. Therefore, water moves across the endodermis into the medulla by osmosis.

Water cannot return to the cortex through the apoplast pathway as this is blocked by the Casparian strip. Therefore, pressure builds up in the cortex, which pushes the water up the xylem.

Root pressure is the pressure created by the action of the endodermis.

Typical mistake

Candidates sometimes confuse the movement of water due to hydrostatic pressure differences with the movement between cells caused by a water potential gradient. Water potential gradients cannot move water all the way up the xylem.

Adhesion

Adhesion is the attraction between the water molecules and the walls of the xylem vessel. It results in the water creeping up the xylem in a process called capillary action.

Transpirational pull

Loss of water vapour from the leaves must be replaced by water in the xylem. As water moves out of the xylem, it creates a pull on the column of water in the xylem. As water is cohesive (the molecules attract one another), the column of water is put under tension and pulled up the stem. This is known as the **cohesion-tension theory**.

Adhesion is the attraction between water molecules and the walls of the xylem.

The **cohesion-tension theory** accounts for the movement of water up the xylem.

Typical mistake

Some candidates state that the water moves up the xylem by cohesion tension. This is incorrect — water moves up the xylem as a result of tension created by the loss of water in the leaves, which draws the whole column of water up the xylem due to the cohesion between the water molecules.

Revision activity

Create a mind map to link together the ideas about how water moves up the xylem.

Adaptations of plants to the availability of water

Xerophytes

REVISED

Xerophytes are plants that are adapted to living in dry (or arid) places. The following adaptations help them to reduce loss of water vapour:

- thick waxy cuticle on the leaves
- smaller leaf area
- stomata in pits
- hairy leaves
- rolled leaves

Revision activity

Draw a leaf from a xerophyte and annotate it with all the features that help the plant conserve water. For each feature, explain how it reduces water loss using the terms *water potential* and *water potential gradient*.

Hydrophytes

REVISED

Hydrophytes are plants that are adapted to living in water, such as water lilies. The following adaptations help them to do this:

1 Leaves and leaf stems have large air spaces (to help them float).
2 The stomata may be on the upper surface of the leaf (to gain carbon dioxide from the air).
3 The stem may be hollow (to allow gases to move to the roots easily).

The mechanism of translocation

Defining translocation

REVISED

Translocation is the movement of **assimilates** (mostly **sucrose**) around the plant. It occurs in the sieve tubes, but the companion cells are important in actively loading assimilates into the sieve tubes.

Translocation is achieved by mass flow. It is caused by creating a high hydrostatic pressure at the source and a lower hydrostatic pressure at the sink. The fluid in the phloem sieve tube then moves from high to low pressure, i.e. down its pressure gradient.

Translocation is an energy-requiring process for transporting assimilates around a plant.

Sources

A **source** is a part of the plant that has a supply of assimilates that are loaded into the phloem. This could be:

- a leaf that has made sucrose from the products of photosynthesis during the spring and summer
- a root that has stored starch and can convert this to sucrose, which happens particularly in spring
- any other storage organ where the plant has stored starch

A **source** is a tissue or organ that supplies assimilates to the phloem.

Sinks

A **sink** is a part of the plant that removes sucrose from the phloem and uses or stores it. This could be:

- the buds or stem tips where growth occurs and energy is needed
- the leaves in spring as they grow and unfold
- the roots in summer and autumn when the plant is storing sugars as starch
- any other organ where the plant may store starch

A **sink** is a tissue or organ that removes assimilates from the phloem and uses them.

Now test yourself

TESTED

5 Explain how a leaf can be a source or a sink.

Answer on p. 225

Creating high pressure at the source

This process is called **active loading**. Sucrose is moved into the sieve tube by a complex process involving active transport:

1 Hydrogen ions are pumped actively out of the companion cells, which uses ATP as a source of energy.
2 The hydrogen ions diffuse back into the companion cells through special co-transporter proteins carrying sucrose molecules into the companion cells.

Exam tip

The details of the mechanism of translocation are described here in some detail, which is needed to achieve full marks.

3 The sucrose builds up in the companion cells and diffuses into the sieve tube through the many plasmodesmata.
4 The water potential in the sieve tube is reduced.
5 Water flows into the sieve tube by osmosis, increasing the pressure.

Creating lower pressure at the sink

As sucrose is used in respiration or converted to starch in the cells of the sink, the concentration decreases. This creates a concentration gradient between the sieve tubes and the cells in the sink. Sucrose diffuses out of the sieve tubes into the cells and the water potential in the sieve tube increases. Water then moves out of the sieve tube by osmosis.

Revision activity

Draw a flow chart to show how the sieve tube elements are actively loaded.

Revision activity

Write a list of all the key terms used in this chapter, then add the meaning of each key term.

Exam practice

1 Which row in the following table correctly identifies the definitions of the terms given? [1]

Row	Plasmodesmata	Symplast pathway	Casparian strip	Transpiration
A	A connection between cells	Water passes through the cytoplasm	A waterproof layer in the endodermis	Loss of water from the leaves
B	A connection between cells	Water passes along the cell walls	A waterproof layer in the endodermis	Loss of water vapour from the leaves
C	A connection between cells	Water passes through the cytoplasm	A waterproof layer in the endodermis	Loss of water vapour from the leaves
D	A connection between cells	Water passes through the cytoplasm	A waterproof layer in the xylem	Loss of water from the leaves

2 (a) Define the term *transpiration*. [2]
(b) Explain why transpiration is an inevitable result of photosynthesis. [2]
(c) Many xerophytes have a thick waxy cuticle and roll their leaves. Explain how these features reduce transpiration. [5]
3 (a) Define the terms *source* and *sink*. [3]
(b) Name two possible sources. [2]
(c) Explain how a leaf can be a sink and a source at different times. [2]
4 (a) Describe the features of the sieve tubes that help mass flow to occur. [3]
(b) Describe and explain how the companion cells are specialised to their role in loading assimilates into the sieve tubes. [3]

Answers and quick quiz 10 online

ONLINE

Summary

By the end of this chapter you should be able to:

- Explain why large plants need a transport system.
- Describe the distribution of xylem and phloem.
- Describe the structure of xylem and phloem and how they are adapted.
- Define transpiration, describe the factors that affect the rate of transpiration and how to measure the rate using a potometer.
- Explain, using the term *water potential*, how water moves between cells and across plant tissues.
- Describe and explain how water moves up the xylem.
- Describe and explain how assimilates are moved in the phloem.

11 Communicable diseases, disease prevention and the immune system

Pathogens and their transmission

The different types of pathogen

REVISED

Pathogens are microorganisms that cause disease. There is a wide range of them and they are transmitted in a variety of ways (Table 11.1). The means of transmission can be categorised as **direct** or **indirect**.

A **vector** is an organism that carries the pathogen from one host to another.

Table 11.1 Selected pathogens and their means of transmission

Name of disease	Organism that causes disease	Means of transmission
Athlete's foot (in humans)	Fungus: *Trichophyton rubrum*	Direct contact with **spores** on the skin surface or on other surfaces
Bacterial meningitis	Bacteria: *Neisseria meningitidis* or *Streptococcus pneumonia*	Direct contact with saliva
Black sigatoka (in bananas)	Fungus: *Mycosphaerella fijiensis*	Indirect transmission through spores spread by the wind
Blight (in potatoes and tomatoes)	Fungus-like organism: *Phytophthora infestans*	Direct contact between infected and uninfected seed potatoes; also indirect by spores in the wind
HIV/AIDS	Virus: human immunodeficiency virus	Direct transmission from an infected person by blood to blood contact; infected needles being shared or reused; infected and unsterilised surgical instruments; accidental needle stick; in semen or vaginal fluid during unprotected sexual intercourse; from mother to baby during birth or breast-feeding
Influenza	Virus: from family Orthomyxoviridae — flu viruses	Indirect transmission by droplet infection
Malaria	Protoctistan: *Plasmodium falciparum*, *P. vivax*, *P. ovale*, *P. malariae*	Indirect transmission via a vector — the vector is the female *Anopheles* mosquito
Ring rot (in potatoes and tomatoes)	Bacteria: *Clavibacter michiganensis* subsp. *sepedonicus*	Direct contact between infected and uninfected potatoes
Ringworm (in cattle)	Fungus: *Trichophyton verrucosum*	Direct contact with spores on the skin surface or on other surfaces
Tobacco mosaic virus	Virus: tobacco mosaic virus	Direct contact between infected and uninfected leaves; also indirect by aphids acting as **vectors**
Tuberculosis (TB)	Bacteria: *Mycobacterium tuberculosis* and *M. bovis*	Indirect transmission by bacteria carried in tiny water droplets in the air

Now test yourself

TESTED

1 Explain how a vector can transmit a pathogen without being infected by the pathogen itself.

Answer on p. 225

Plant defences against pathogens

REVISED

Organisms are surrounded by pathogens and have evolved defences against them. Plant defences can be divided into physical and chemical.

Physical defences

Cellulose cell walls, lignin thickening of cell walls, waxy cuticles and bark all help to prevent entry of a pathogen. Inside the plant, old vascular tissue is blocked to prevent a pathogen spreading easily. **Callose** is deposited around the sieve plates in older sieve tubes and blocks flow. Tylose (a balloon-like swelling) also blocks old xylem vessels.

Callose is a large polysacchride that blocks old sieve tubes.

When a pathogen is detected, the following barriers can be enhanced to prevent it spreading through the plant:

1 The stomata close to prevent entry to the leaves.
2 Cell walls are thickened with additional cellulose.
3 Callose is deposited between the plant cell wall and cell membrane near the invading pathogen. This strengthens the cell wall and blocks the plasmodesmata.
4 Necrosis (deliberate cell 'suicide') — cells surrounding the infection are killed to reduce access to water and nutrients. Necrosis is caused by intracellular enzymes that are activated by injury. These enzymes kill damaged cells and produce brown spots on leaves.
5 Canker — necrosis of woody tissue in the main stem or branch.

Now test yourself

2 Explain why blocking the phloem with callose and the xylem with tylose reduces the spread of a pathogen.

Answer on p. 225

TESTED

Chemical defences

Chemicals such as terpenes and tannins are present to prevent entry of a pathogen. However, other chemicals can be released when a pathogen is detected. Chemical defences include:

1 terpenoids — essential oils with antibacterial and antifungal properties, e.g. menthols produced by mint plants
2 phenols (e.g. tannin) — found in bark, which has antibiotic and antifungal properties. Tannins inhibit insect attack by interfering with their digestion
3 alkaloids (e.g. caffeine, nicotine, cocaine and morphine) — give a bitter taste to inhibit herbivores feeding. They also interfere with metabolism by inhibiting or activating enzyme action
4 defensins — small cysteine-rich proteins that inhibit ion transport channels in the plasma membrane of pathogen cells
5 hydrolytic enzymes — found in the spaces between cells. These include chitinases, which break down the chitin in fungal cell walls; glucanases, which hydrolyse the glycosidic bonds in glucans of bacterial walls; and lysosymes, which degrade bacterial cell walls.

Animal defences against pathogens

REVISED

Animal defences against pathogens take two forms:

1. **primary defences**, which prevent entry of pathogens into the body
2. **secondary defences**, which combat pathogens that have already entered the body

Primary defences prevent the entry of the pathogen into the body.

Secondary defences help to remove a pathogen after it has entered the body.

Primary defences

The skin is the main primary defence against pathogens and parasites. It provides a barrier to the entry of microorganisms.

Blood clotting and skin repair

Damage to the **skin** opens the body to infection. **Blood clots** reduce the loss of blood and make a temporary seal, preventing access by pathogens. Blood-clotting involves calcium ions and at least 12 other clotting factors that are released by platelets or from the damaged tissue. These factors activate an enzyme cascade, which produces insoluble fibres (Figure 11.1). The clot dries out to form a scab. Over time, the scab shrinks, drawing the sides of the cut together. Fibrous collagen is deposited under the scab and stem cells in the epidermis divide to form new cells. These cells differentiate to form new skin.

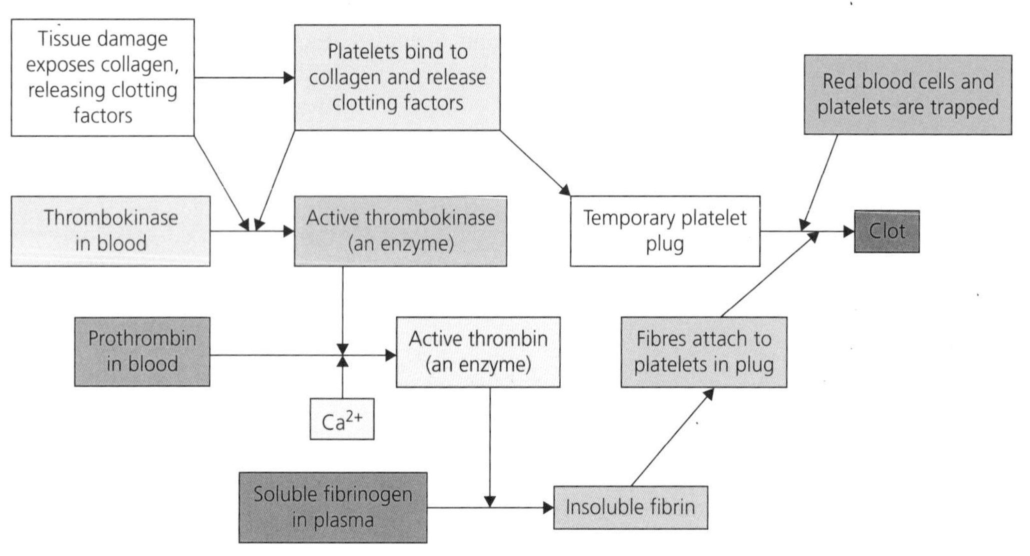

Figure 11.1 An enzyme cascade causes blood to clot

Inflammation

Inflammation is the swelling and redness seen in infected tissues. Infected tissue often feels hot and tender. This is caused by a **cell-signalling** substance called histamine. Histamine is released from mast cells and it has a range of effects that act to help combat the infection. It causes vasodilation and makes the capillary walls more permeable to white blood cells and some proteins. More plasma enters the tissue fluid, causing oedema (swelling). The excess tissue fluid is drained into the lymphatic system, moving the infecting pathogens towards the lymph nodes where lymphocytes can initiate the specific **immune response**.

Now test yourself

3 Explain why the blood-clotting process needs to be so complex.

Answer on p. 225

TESTED

The **immune response** is the body's response to invasion by pathogens.

Mucous membranes

Any areas where the skin is incomplete are protected by **mucous membranes**. This includes the airways, lungs, digestive system, ears and genital areas. The epithelial layer contains mucus-secreting goblet cells. The mucus traps any pathogens, immobilising them. Some areas, such as the airways, also have ciliated cells. Cilia are tiny, hair-like organelles that can move in a coordinated fashion to waft the layer of mucus along.

Mucous membranes are sensitive to irritation. They respond to the presence of microorganisms or the toxins they release. This causes expulsive reflexes (e.g. coughing, sneezing and vomiting). In a cough or sneeze, the sudden expulsion of air carries with it the microorganisms causing the irritation.

Antigens are molecules on the surface of cells that the immune system can use to recognise pathogens.

Antibodies are proteins that are secreted in response to stimulation by the appropriate antigen. They have specific binding sites and are capable of acting against the pathogen.

Secondary defences

The non-specific immune response

Phagocytosis

This involves phagocytic white blood cells (**neutrophils**) that engulf and destroy any non-self cells. Neutrophils are cells that contain a lobed nucleus and dense cytoplasm that contains many **lysosomes** and mitochondria. They also have a well-developed cytoskeleton.

Non-self cells are recognised because they have proteins on their cell-surface membranes called **antigens**. They may also be identified by the presence of opsonins, which are non-specific **antibodies** that bind to pathogens and act as binding sites for the phagocyte.

Phagocytosis follows a particular sequence (Figure 11.2):

1 The bacteria are engulfed by the neutrophil.
2 They are surrounded by a vacuole called a **phagosome**.
3 Lysosomes fuse to the phagosome.
4 Lytic enzymes are released into the phagosome.
5 The bacteria are hydrolysed (digested) and the nutrients can be reabsorbed into the cell.

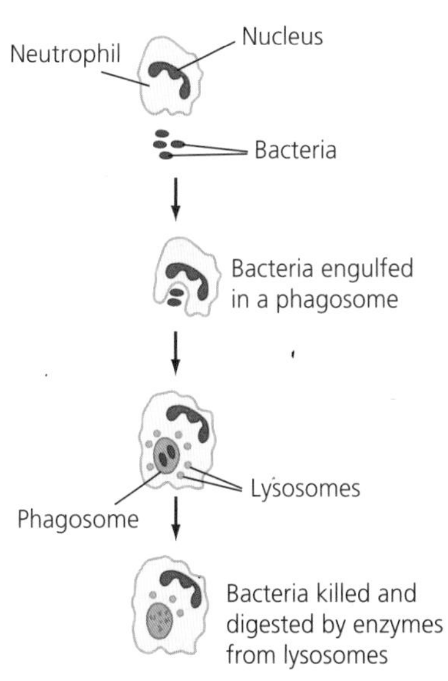

Figure 11.2 The stages involved in phagocytosis

Antigen presentation

Some phagocytes (macrophages) can engulf the pathogen and process it so that the antigen is kept whole. This antigen can then be placed in a special protein complex on the cell-surface membrane. This is known as **antigen presentation**. These **antigen-presenting cells** can then be used to initiate the **specific immune response**.

Antigen presentation involves placing an antigen on the cell-surface membrane of a phagocytic cell.

The **specific immune response** means that the response deals only with one pathogen that possesses one particular antigen.

The specific immune response

The specific response relies on the action of the **B lymphocytes** and **T lymphocytes**, which are cells with a large nucleus. They have specific receptors on their cell-surface membranes. These receptors show a wide diversity of shapes, each of which is complementary to the shape of an antigen on a specific pathogen.

The specific response is more complex than the non-specific response and involves a series of stages (Figure 11.3).

1 *Antigen presentation.*
2 *Clonal selection.* There may be only a few lymphocytes that carry the correct receptors to bind with the specific antigen. Normally, there is a B lymphocyte, a **T helper** lymphocyte and a **T killer** lymphocyte. These cells must be found and activated. They are selected and activated by coming into contact with the specific foreign antigen. This ends the cell-signalling role of the antigen presenting cell. The action of macrophages makes this selection more likely.

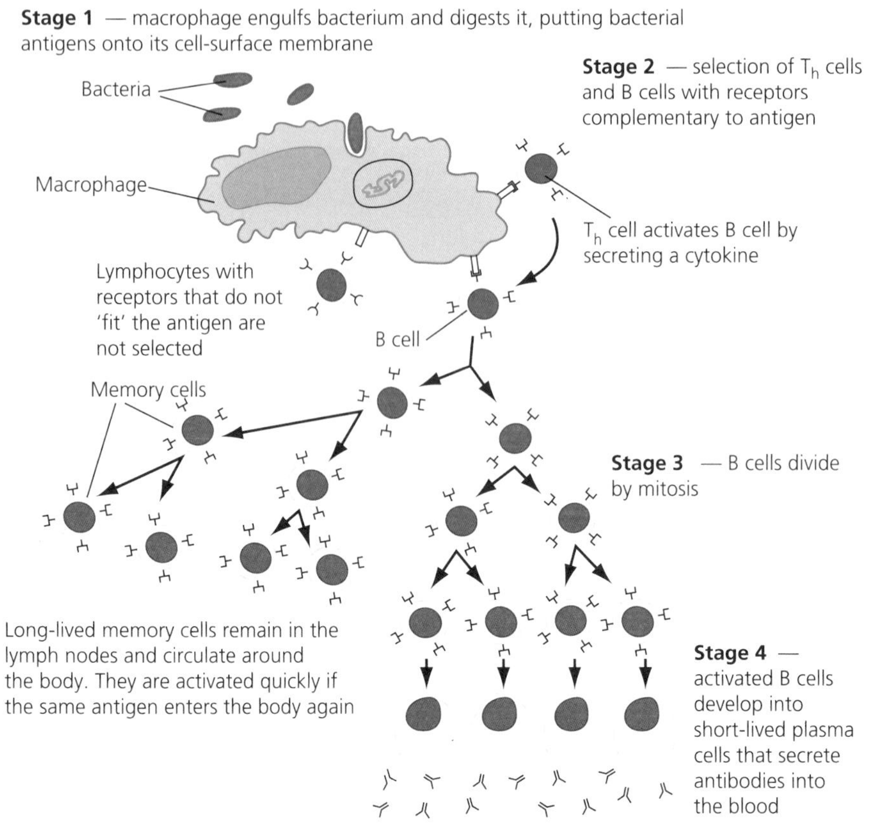

Stage 1 — macrophage engulfs bacterium and digests it, putting bacterial antigens onto its cell-surface membrane

Bacteria

Macrophage

Stage 2 — selection of T_h cells and B cells with receptors complementary to antigen

T_h cell activates B cell by secreting a cytokine

Lymphocytes with receptors that do not 'fit' the antigen are not selected

B cell

Memory cells

Stage 3 — B cells divide by mitosis

Long-lived memory cells remain in the lymph nodes and circulate around the body. They are activated quickly if the same antigen enters the body again

Stage 4 — activated B cells develop into short-lived plasma cells that secrete antibodies into the blood

Figure 11.3 The stages of the specific immune response

3 *Clonal expansion and differentiation.* Once activated, the specific lymphocytes increase in numbers by mitosis. There will normally be a clone of B lymphocytes, a clone of T helper lymphocytes and a clone of T killer lymphocytes. Cells from the clone of B lymphocytes will differentiate into plasma cells and **memory cells**. The plasma cells are short-lived. They produce and release the antibodies which are proteins that have specific binding sites and are capable of acting against the pathogen.
4 *Differentiation.* Cells from the T helper lymphocyte clone will differentiate into T helper cells and memory cells. The T helper cells secrete hormone-like substances called **cytokines** and **interleukins** that help to stimulate the B lymphocytes and the macrophages. Memory cells are long-lived and remain in the blood for some time, providing immunological memory or long-term immunity. If the same pathogen invades again, it will be recognised and attacked more quickly. The T killer clones manufacture harmful substances and attack and kill host cells that are already infected. **T regulator** cells have a role in closing down the immune response once the pathogen has been removed.

Typical mistake

Many candidates still seem to confuse antigens and antibodies.

Exam tip

To gain full marks, you must be specific — antibodies are made by the plasma cells.

Revision activity

Try to visualise this whole process by drawing a simple flow diagram.

Now test yourself

TESTED

4 Explain why pathogens have antigens that allow them to be recognised as foreign on their surface.

Answer on p. 225

Primary and secondary immune responses

Primary responses

REVISED

The first time a pathogen invades the body, it produces a **primary response.** This takes a few days as the specific B and T lymphocytes must be selected, the cells must divide and then differentiate and the antibodies must be manufactured. As a result, the peak of activity and the maximum concentration of antibodies are not achieved until several days after infection.

The **primary response** is the immune system's response to a first infection.

Secondary responses

REVISED

On any subsequent invasion by the same pathogen with identical antigens on its surface, the immune response is a **secondary response**. The immune system can respond much more quickly and with higher intensity — it is much more effective. This is because the blood carries many memory cells that are specific to this pathogen. They divide and differentiate into plasma cells, which then manufacture antibodies. The response is quick enough to prevent the pathogen taking hold and causing symptoms of illness.

The **secondary response** is the immune system's response to a second or subsequent infection by the same pathogen.

The structure and general functions of antibodies

The general structure

REVISED

Antibodies are large globular proteins. Most antibodies are similar in structure and possess the same basic components. They consist of four polypeptides held together by disulfide bridges (Figure 11.4).

One end of an antibody is known as the constant region. This has a binding site that can be recognised by phagocytes. It may help binding of phagocytes. The opposite end is known as the variable region. This has binding sites that are specific to a particular antigen. They have a shape that is complementary to the shape of the antigen and can bind to that antigen. The antibody molecule also has a hinge region that allows some flexibility to enhance binding to more than one pathogen.

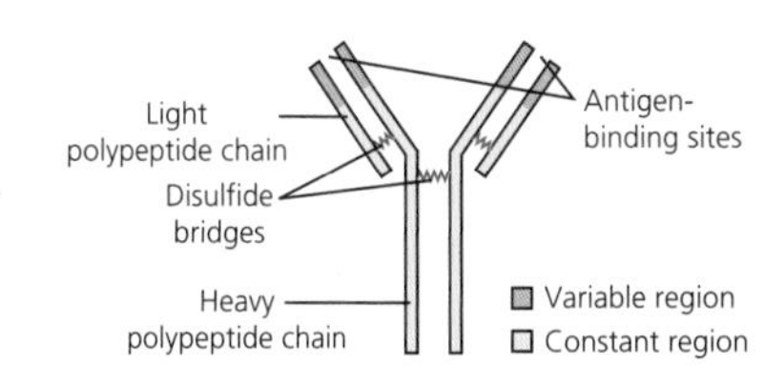

Figure 11.4 The structure of an antibody molecule

The action of antibodies

REVISED

Antibodies can act on pathogens in a number of ways. These include:

1. **opsonins** — antibodies that bind to a pathogen and help phagocytes to bind
2. **agglutins** — antibodies that bind to two or more pathogens so that they are held together. This prevents them entering cells and reproducing. Some antibodies have many binding sites. These can be used to bind to a number of pathogens.
3. **anti-toxins** — antibodies that bind to the toxin, making it harmless

Types of immunity

Active and passive immunity

REVISED

Active immunity is acquired through activation of the immune system. It involves the selection of specific lymphocytes and the production of antibodies and memory cells. The memory cells remain in the blood for a long time, providing long lasting immunity.

Passive immunity is acquired from another source. Antibodies may be injected straight into the blood or acquired from a mother's milk. These antibodies do not last long, but they do give immunity from a specific pathogen for a period of time. No memory cells are made, so the immunity is not permanent.

Active immunity is immunity acquired by activation of the immune system.

Passive immunity is when someone is given antibodies produced by someone else.

Natural and artificial immunity

REVISED

Natural immunity is acquired in the course of everyday activity. Natural active immunity may be the result of catching a flu virus from someone who sneezes. Natural passive immunity may be acquired through breast milk or the mother's placenta.

Artificial immunity is acquired by human intervention. Artificial active immunity may be the result of a vaccination. Artificial passive immunity results from an injection of antibodies.

Autoimmune diseases

Defining autoimmune diseases

REVISED

An **autoimmune disease** is one in which the immune system attacks the body's own healthy cells and tissues. The B and T lymphocytes usually respond to antigens on harmful organisms such as bacteria or viruses. In an autoimmune disease, the lymphocytes do not distinguish between these 'foreign' antigens and your own. As a result, they release antibodies that attack your own tissues.

This attack may be due to the exposure of antigens that are not normally exposed, such as certain molecules usually found only in the nucleus. The exact cause of autoimmune disease is not known, but it may be due to changes triggered by drugs or by an infection with certain bacteria or viruses. It also seems that certain people are more genetically prone to autoimmune disease.

An autoimmune disease may cause:

1. damage to body tissues and organs
2. abnormal growth of an organ
3. changes in organ function

Examples include **arthritis** (painful inflammation of joints) and **lupus**, which can affect any part of the body causing swelling and pain.

The principles of vaccination

Routine vaccinations

REVISED

Vaccination is the deliberate introduction of antigenic material in order to stimulate the production of antibodies. Antigenic material can be:

- whole live microorganisms (this is more effective than using dead ones as the live organisms can reproduce and mimic an infection better)

- dead microorganisms
- attenuated (weakened) organisms
- a surface preparation of antigens
- a toxoid (a harmless form of a toxin)

Herd and ring vaccination

REVISED

In order to be effective, vaccinations need to be used appropriately. Herd vaccination involves the systematic vaccination of all or most members of a population. This prevents the pathogen being transmitted from person to person. Ring vaccination is a response to an outbreak. All the people in the area surrounding the outbreak are vaccinated in order to prevent transmission and to isolate the outbreak in one area.

Now test yourself

5 Explain why using a live organism in a vaccination is better than injecting antibodies or antigens.

Answer on p. 225

TESTED

Possible sources of medicines

New medicines

REVISED

Scientists are always looking for new medicines to help combat disease. **Microorganisms** and **plants** produce a wide range of molecules that may be of benefit in fighting disease. Among the huge diversity of plants and microorganisms in the natural world there may be organisms that produce chemicals that are beneficial against a wide range of pathogens or in fighting diseases such as cancer. We just need to find them.

Every species of organism that is allowed to become extinct could potentially hold the cure for a major disease. It is therefore important that we try to maintain biodiversity and conserve as many species as possible, just in case the molecules they are capable of producing may prove to be useful.

Personalised medicine

Screening the genomes of many plants and microorganisms enables scientists to identify compounds that may have a medicinal value. As sequencing technology and molecular modelling techniques improve, it may be possible to sequence the DNA of an individual to assess a specific genetic disorder and develop a treatment personalised to the individual, such as an individualised combination of drugs. Such drugs could be those identified from plants or microorganisms, or these compounds could be modified to become more effective. In the future, it may even become possible to design a drug that is specific to the needs of one individual.

Synthetic biology

Scientists can modify natural molecules for use as medicines. Synthetic biology is the development of new molecules or systems. This could be making new enzymes to create a new product or using natural enzymes to produce a new effect. For example, bacteria are being genetically modified for use as biosensors. Scientists have created new pathways that enable bioluminescent bacteria to release light in response to specific pollutants.

Benefits and risks of antibiotics

The wide use of antibiotics

REVISED

The first antibiotic, **penicillin**, was discovered by Alexander Fleming in 1928. Since then, many different antibiotics have been discovered and used widely to treat bacterial infections.

The use of antibiotics became widespread during the Second World War to prevent the infection of wounds, which has saved many lives. However, their misuse has enabled microorganisms to develop resistance. ***Clostridium difficile*** (*C. diff*) and methicillin-resistant *Staphylococcus aureus* (**MRSA**) are well known for their multiple **resistance** to antibiotics.

Exam practice

1 Which row in the following table shows the correct definitions of the terms given? [1]

Row	Antigen	Antibody	Pathogen	Vaccination
A	A protein made by the immune system	A molecule on the surface of a pathogen	A microorganism that causes disease	The introduction of antigenic material
B	A molecule on the surface of a pathogen	A protein made by the immune system	A microorganism that causes disease	The introduction of antigenic material
C	A molecule on the surface of a pathogen	A protein made by the immune system	An organism that lives on another orgainism	The introduction of antigenic material
D	A molecule on the surface of a pathogen	A protein made by the immune system	A microorganism that causes disease	An injection

2 (a) Describe three ways in which HIV is transmitted. [3]

(b) The following table shows the percentage of people with new HIV infections in four regions.

	Percentage of people with new HIV infections			
Year	Western Europe	Eastern Europe	Far East	Sub-Saharan Africa
1980	0.0	0.0	0.0	0.0
1990	0.1	0.2	0.1	2.0
2000	0.4	0.5	0.3	8.5
2010	0.3	0.9	0.3	15.2

(i) Describe the trends shown by the data in the table. [4]

(ii) Suggest why the trend in Western Europe is different from that in other areas. [3]

(c) Discuss the advantages and disadvantages of treating HIV/AIDS patients with antibiotics. [4]

3 (a) Explain the difference between antigens and antibodies. [3]

(b) Describe the structure of an antibody and explain how it is adapted to its function. [4]

(c) Suggest and explain why people in underdeveloped parts of the world such as sub-Saharan Africa may be unable to make sufficient antibodies. [4]

Answers and quick quiz 11 online

ONLINE

Summary

By the end of this chapter you should be able to:

- Describe the types of pathogen that cause disease and how they are transmitted.
- Describe plant defences against pathogens.
- Describe the primary defences against pathogens in animals.
- Describe the action of phagocytes.
- Describe the structure and action of B and T lymphocytes.
- Compare and contrast primary and secondary immune responses.
- Describe the structure and action of antibodies.
- Compare and contrast active, passive, natural and artificial immunity.
- Explain how vaccination can control disease.
- Outline the possible sources of new medicines.
- Outline the benefits and risks of antibiotics.

12 Biodiversity

The levels of biodiversity

Exam tip

This topic covers key scientific terms with precise meanings. You should learn them and use them only in the correct context.

Considering biodiversity at different levels

REVISED

Biodiversity is the variety of life. It includes all the different plant, animal, fungus and microorganism species in the world, the genes they contain and the ecosystems of which they form a part. Biodiversity can be considered at three levels: the range of **habitats** within an ecosystem, the range of **species** within a habitat and the **genetic** variation within a species (different breeds within a species).

Biodiversity is the variety of life on Earth.

A **habitat** is the home or environment of an animal, plant or organism.

A **species** is a group of organisms with similar adaptations that live and breed together to produce fertile offspring.

Revision activity

Draw a mind map with biodiversity in the centre. Include the three levels of biodiversity and explain what they mean.

Sampling

REVISED

Most habitats are large in area and have large numbers of plants and animals. It is therefore impossible to count the number of individuals in each species. **Sampling** involves studying small parts of the habitat in detail and then multiplying up to calculate the size of a population. It is assumed that the sample plots are representative of the entire habitat.

When sampling, it is important to consider the:

- size of the samples — this depends on the size of the habitat
- number of sample areas used — the more sample areas used the better, as the results will be more reliable
- sampling technique used — this must be identical in every sample

The sampling should not disturb the habitat more than is essential.

Techniques

There are two main techniques for sampling: random and non-random.

Random sampling avoids bias. It can be achieved by using a computer to generate random numbers, which are then used as coordinates to locate sample areas on an imaginary grid placed over the habitat.

Non-random sampling includes:

1 **Opportunistic sampling**, which involves using prior knowledge to select sample sites or changing the sampling strategy once onsite.
2 **Stratified sampling**, which involves carrying out samples in each recognisable sub-habitat.
3 **Systematic sampling**, which involves carrying out sampling at fixed intervals in each direction.

Now test yourself

1 Explain why it is important to avoid bias in sampling.

Answer on p. 225

TESTED

Methods

The method of sampling used depends on the type of vegetation in the habitat and what type of organisms are being examined. It is important to measure both the number of species (species richness) and the number of individuals in each species (species evenness).

Plants

Large plants such as trees can be counted individually. Smaller plants can be sampled using quadrats. These are square frames of a suitable size that are placed over a random site and examined closely to identify all the plants inside the quadrat. A quantitative sample can be achieved by measuring the percentage cover of each species within the quadrat. This can be done using the following methods:

- point sampling — place a point frame in the quadrat and count the number of examples of each species that touch each point
- grid sampling — divide the quadrat using string into a known number of smaller squares (often 100) and then estimate how many squares are occupied by each species

Animals

Large animals can be sampled by careful observation and counting. Smaller animals will need to be caught or trapped.

Small mammals can be trapped using a humane trap such as a Longworth trap. It is possible to estimate the population size by the mark and recapture technique. This involves two separate trapping sessions. The animals caught the first time are marked in a way that causes them no harm. If the number of animals trapped in the first session is T1, the number caught in the second session is T2 and the number caught in the second session that are already marked is T3, the total population can be given by the formula:

$$\text{Number in population} = T1 \times \frac{T2}{T3}$$

Ground-living invertebrates can be collected using a pitfall trap, whereas invertebrates in leaf litter can be collected using a Tullgren funnel. Invertebrates in trees can be collected by using a stick to knock a branch and collecting in a sheet the organisms that fall to the ground. Invertebrates in grass and shrubs can be collected by sweep netting, and pond life can be sampled by netting.

Species richness and species evenness

Measuring species richness and species diversity

REVISED

A count of the number of species in a habitat is called the **species richness**. However, a habitat that is dominated by just one species with only one or two individuals of each of the other species would not be considered to be biologically diverse. A habitat in which all species are equally represented is more diverse. This is known as **species evenness**. Therefore, a measurement of the biodiversity of a habitat should account

Species richness is the number of different species in a habitat.

Species evenness is how evenly each species is represented throughout a habitat.

for both the number of species and the number of individuals in each species living in that habitat. A diverse habitat would contain a large number of species, all of them represented by a sizeable population rather than by just one or two individuals.

Now test yourself

TESTED

2 Explain the significance of species richness and species evenness.

Answer on p. 225

Simpson's Index of Diversity

Understanding and using the formula

REVISED

Simpson's Index of Diversity measures the biodiversity of a habitat. There are several versions of the formula, so make sure you are consistent. The most commonly used version is:

$$D = 1 - (\Sigma(n/N)^2)$$

where n = the total number of individuals in a particular species and N = the total number of individuals in all species.

If it is not possible to count all the individual plants in an area, percentage cover can be used. The resultant value always ranges between 0 and 1.

Worked example

The data collected from a field is recorded as % cover in columns one and two. The processing used to calculate Simpson's Index of Diversity is shown in columns three and four.

Table 12.1 **Calculations for Simpson's Index of Diversity**

Species	Percentage cover	n/N	$(n/N)^2$
Yorkshire fog	16	0.16	0.0256
Meadow grass	74	0.74	0.5476
Bent grass	2	0.02	0.0004
Thistle	3	0.03	0.0009
Buttercup	4	0.04	0.0016
Dock	1	0.01	0.0001
Total	100	1.00	Σ = 0.5762

Typical mistake

Some candidates don't find calculations easy — here they often forget to square each number or to subtract from 1 as the final step in the calculation.

Therefore, using Simpson's Index of Diversity:

$$D = 1 - (\Sigma(n/N)^2)$$

$$D = 1 - (0.0256 + 0.5476 + 0.0004 + 0.0009 + 0.0016 + 0.0001)$$

$$= 1 - 0.5762$$

$$= 0.42$$

Exam tip

You may be asked to calculate Simpson's Index of Diversity from data provided. Calculations involving the index are not as difficult as many students imagine. The formula and a partly completed table are often provided.

Interpretations of high and low values

A **high value** (close to 1) indicates:

- a habitat with high diversity
- there are a good number of species (high species richness)
- the species are relatively evenly represented (high species evenness)
- the habitat should be stable and may survive some disruption
- the habitat is probably one that is worth conserving

A **low value** (close to 0) indicates:

- a habitat with low diversity
- the habitat is dominated by one or a few species
- the habitat may be unstable and damaged by disruption
- the habitat may be manmade

In the worked example above, a diversity index of 0.42 is not particularly high. This habitat is dominated by one species (meadow grass). If this species were harmed by human action, the habitat may be unstable.

Now test yourself

3 Explain why an ecologist might want to monitor the diversity of a habitat.
4 Explain why a conservationist may want to conserve a more diverse habitat.

Answers on pp. 225–26

TESTED

Genetic biodiversity

Assessing genetic diversity

REVISED

The **genetic diversity** of a population may be important to conservationists wishing to maintain the health of a captive population in a zoo or rare-breed centre. It may also be important in pedigree charts. Genetic diversity is increased when there is more than one **gene variant** (**allele**) for a particular **locus**. If there are several alleles, it is called a **polymorphic gene locus**. A measure of genetic diversity is given by the formula:

$$\text{Proportion of polymorphic gene loci} = \frac{\text{number of polymorphic gene loci}}{\text{total number of loci}}$$

The factors affecting biodiversity

REVISED

Many species have become rare or endangered, often because of human activity. The main factors that affect biodiversity include:

1. the rapid **human population growth**, which means that more space and resources are taken up to supply living space and food
2. the increasing use of **agriculture** (**monoculture**) as an efficient way to produce food and other products. This decreases biodiversity and reduces the size of natural habitats, which may become unstable as a result
3. the increasing release of waste products that pollute the atmosphere, causing **climate change**

Maintaining biodiversity

The reasons for maintaining biodiversity

REVISED

It is important to conserve endangered species for a variety of reasons, many of which can be applied generally to most endangered species.

A **keystone species** is one that has a disproportionate effect on the ecosystem — the loss of one species may affect many others.

Ecological reasons

Many species are in decline because of habitat destruction — finely balanced ecosystems may be disrupted by a small change. This is particularly true where **keystone species** are involved. For example, beavers make dams that cause flooding and introduce new conditions in which many aquatic species live. All those species would be lost if beavers were not present.

Exam tip

The reasons for conservation are generic, but questions in the examination are likely to ask about a specific case. Be ready to apply these generic ideas to the case under investigation.

The whole balance of life on Earth is maintained by the activity of species within ecosystems and we do not know what knock-on effects the loss of one species may have. The **ecological reasons** for maintaining the correct functioning of ecosystems include:

- fixing of energy from sunlight
- regulation of the oxygen and carbon dioxide levels in the atmosphere
- fresh water purification and retention
- soil formation
- maintenance of soil fertility
- mineral recycling
- waste detoxification and recycling

Economic reasons

Economic reasons for conserving species include:

- Growth of food and timber relies on the correct functioning of ecosystems. Even the soil relies on the ecosystem functioning correctly. If an ecosystem is disrupted, the effects may be far-reaching. An unbalanced ecosystem could cause **soil depletion** so that it loses fertility. This is particularly obvious where monoculture is used constantly.
- Pollination of many crops relies on insects, particularly bees.
- Natural predators to pests reduce the need for pesticides.
- As yet unknown species may contain molecules that are effective medicines.

Typical mistake

Many students write long, heartfelt responses about the right of all organisms to live and how humankind should not 'play God'. This sort of response may gain some credit, but no more than 1 or 2 marks.

Aesthetic reasons

Aesthetic reasons for maintaining biodiversity are important:

- Everyone enjoys nature or its benefits in one way or another. A healthy, well-balanced ecosystem with its variety of life forms, colours and activity is complex and beautiful.
- Being surrounded by natural systems relieves stress and helps recovery from injury.
- Maintaining the landscape.

In situ and *ex situ* conservation

REVISED

In situ conservation

***In situ* conservation** involves conserving a species in its natural habitat by creating **marine conservation zones** and **wildlife reserves**. Table 12.2 outlines the advantages and disadvantages of this form of conservation.

Table 12.2 The advantages and disadvantages of *in situ* conservation

Advantages	Disadvantages
The organisms are in their normal environment The habitat is conserved, along with all the other species living in it The organisms will behave normally It generates work for local people looking after the reserve Ecological tourism can generate income	It can be difficult to monitor the organisms and ensure that they are healthy The environmental factors that caused the decline in numbers may still be present Poaching or hunting may continue There may be food shortages Disease will be difficult to treat Predators can be difficult to control

Ex situ conservation

***Ex situ* conservation** involves conserving a species using controlled habitats away from its normal environment. **Seed banks**, **botanic gardens** and **zoos** all keep groups of individuals of endangered species. Table 12.3 outlines the advantages and disadvantages of this form of conservation.

Table 12.3 The advantages and disadvantages of *ex situ* conservation

Advantages	Disadvantages
Research is easy Health can be monitored Controlled breeding reduces inbreeding Sharing resource increases genetic diversity Reintroduction to the wild Use less space than botanic gardens	The organisms are living in an unnatural habitat The organisms may not behave as normal The organisms may not breed There is little point in conserving individuals if their natural habitat is lost and there is nowhere for them to return to

Exam tip

The principles of conservation are straightforward. Therefore, questions about conservation are likely to be applied to a specific case. Learn to apply your knowledge to new contexts.

Typical mistake

Many candidates can answer questions on this topic in generic terms but tend to get thrown by specific examples.

Revision activity

Look at a local wildlife park's website to find out what activities are carried out for international conservation.

Now test yourself

5 Explain why it is important to keep more than one population of an endangered species.

Answer on p. 226

TESTED

International and local conservation agreements

REVISED

For conservation activities to be effective, it is essential that all parties agree on what must be done. A wide range of agreements are aimed at ensuring that conservation efforts are successful.

The Convention on International Trade in Endangered Species (CITES)

The **Convention on International Trade in Endangered Species (CITES)** is an international agreement between governments, to which countries adhere voluntarily. Its aim is to ensure that international trade in specimens of wild animals and plants does not threaten their survival.

The Rio Convention on Biological Diversity (CBD)

The **Rio Convention on Biological Diversity (CBD)** recognises that people need to secure resources of food, water and medicines. However, it promotes development that is sustainable and partner countries agree to adopt *ex situ* conservation measures with shared resources.

The Countryside Stewardship Scheme (CSS)

The **Countryside Stewardship Scheme (CSS)** aims to enhance the natural beauty and diversity of the UK's countryside and improve public access. This includes looking after wildlife habitats, retaining the traditional character of the land and protecting historic features and natural resources.

It applies to land that is not considered to be an environmentally sensitive area. Payments are made to landowners to manage the land in a suitable manner and for capital works such as hedge laying, planting and repairing dry-stone walls.

Revision activity

Write a list of all the key terms used in this chapter, then add the meaning of each key term.

Now test yourself

TESTED

6 Explain why international cooperation is essential for successful conservation.

Answer on p. 226

Exam practice

1 (a) State what is meant by the terms *biodiversity*, *species richness* and *species evenness*. [3]

(b) Biodiversity can be measured at three levels: habitat, species and genetic. Explain what is meant by each level and explain the significance of high diversity in each case. [6]

(c) A student collected the following data from two fields.

Species	Percentage cover	
	Field A	Field B
Rye grass	76	14
Bent grass	0	84
Fescue	45	0
Dandelion	9	0
Buttercup	24	2
Daisy	18	0
Total	172	100

(i) Calculate the value of Simpson's Index of Diversity for field B. [2]

(ii) The student calculated Simpson's Index of Diversity for field A. The value was 0.704. Suggest what the relative values for fields A and B mean about their diversity and their value as habitats. [4]

2 The mountain gorilla (*Gorilla beringei beringei*) is an endangered species. There are thought to be about 790 individuals left in the wild. These live in two populations, one in the Virunga mountains of central Africa and the other in southwest Uganda.
   (a) Explain why such a low population puts the species at risk. [3]
   (b) (i) Suggest how local people could help to conserve these gorillas. [3]
      (ii) Suggest how *ex situ* conservation techniques could be used to help the gorillas. [3]
3 Explain why seed banks are considered to be more important than botanical gardens. [2]
4 (a) What do the initials CITES mean? [1]
   (b) (i) Describe the aims of CITES. [2]
      (ii) Suggest two ways in which these aims can be achieved. [2]
   (c) Describe the role of the Countryside Stewardship Scheme. [3]

Answers and quick quiz 12 online

ONLINE

Summary

By the end of this chapter you should be able to:

- Explain how biodiversity can be considered at the level of habitat, species and genetic.
- Explain the importance of sampling when measuring the biodiversity of a habitat.
- Describe how samples can be taken.
- Describe how to measure biodiversity in a habitat.
- Use Simpson's Index of Diversity.
- Outline the significance of high and low values of the Simpson's Index of Diversity.
- Describe how genetic diversity can be estimated.
- Outline the reasons for the loss of biodiversity.
- Describe the conservation of endangered plant and animal species, both *in situ* and *ex situ*, with reference to the advantages and disadvantages of these two approaches.
- Discuss the importance of agreements to the successful conservation of species with reference to CITES, the Rio Convention on Biodiversity and the Countryside Stewardship Scheme.

13 Classification and evolution

Biological classification

Taxonomic hierarchy

REVISED

Biologists use **biological classification of species** to place living things into groups to make them easier to study. These groups are called taxonomic groups or taxa (singular: taxum). The **taxonomic hierarchy** is shown in Figure 13.1. Therefore, **domain** is the largest group (or taxum) and **species** the smallest. Similar species are placed in a **genus**. Similar genera are placed in a **family** etc.

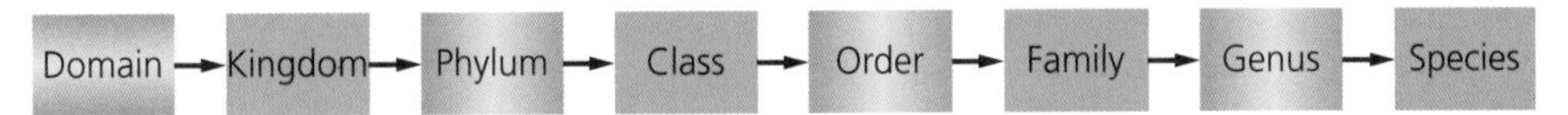

Figure 13.1 The hierarchy of biological classification

Biological classification of species is placing living things into groups.

Now test yourself

1 Explain why the order of taxonomic groups is known as a hierarchy.

Answer on p. 226

TESTED

The binomial system

The system and its advantage

REVISED

The **binomial system** is the way in which we use two Latin words to name each species. The first name is the name of the genus to which the species belongs and the second name is the specific or species name. For example, in the term *Homo sapiens*, *Homo* is the genus of man and *sapiens* is our species name. Both of the Latin names should always be written in italics or underlined. The genus name should be written with an upper case first letter, whereas the species should be in lower case.

The main **advantage** of this system is to avoid the confusion that can arise if local names are used. The binomial system is recognised worldwide.

Typical mistake

Candidates often use the binomial system incorrectly, forgetting to capitalise the genus name or not using italics for the genus and species names.

The features used to classify organisms

The five kingdoms

REVISED

Carl Linnaeus set up the first true system of classification system, which is still used today. He used the organisms' **observable features** and grouped them according to the number of **similarities** present.

More recently, taxonomists study organisms in ever-more detail, describing all of their observable features. They then decide if any differences are simply a variation within the species or significant enough to classify the organism as a different species.

In the **five kingdoms** system of classification, the kingdoms are separated by characteristic features that are summarised in Table 13.1.

Table 13.1 The characteristic features of the five kingdoms

	Kingdom				
Feature	**Prokaryotae**	**Protoctista**	**Fungi**	**Plantae**	**Animalia**
Cellular	Unicellular	Unicellular, some multicellular (algae)	Acellular (body composed of mycelium). Yeasts are unicellular	Multicellular	Multicellular
Nucleus	No	Yes	Yes (cytoplasm is multinucleate)	Yes	Yes
Membrane-bound organelles	No	Yes	Yes	Yes	Yes
Cell wall	Yes (made of peptidoglycan)	Present in many species	Yes (made of chitin)	Yes (made of cellulose)	No
Nutrition	Autotrophic, heterotrophic or parasitic	Autotrophic, heterotrophic or parasitic	Heterotrophic	Autotrophic (photosynthetic)	Heterotrophic
Locomotion	Some have flagella	Some have an undulipodium, some have cilia	None	None	Muscular tissue

Now test yourself

TESTED

2 Explain why fungi were once classified as plants, but have been reclassified in their own kingdom.

Answer on p. 226

New classification systems

REVISED

Similarities in biological molecules

More recent approaches involving sequencing of DNA and proteins have been used to classify organisms. This evidence from **biological molecules** provides a **genetic** and biochemical comparison, which is thought to be more accurate than comparing observable features. As mutations which cause changes in the DNA sequence are random, more mutations (differences in sequence) mean a longer time has elapsed since the species became separate.

The three domains of life

Recent research on the sequence of bases in the RNA of the ribosomes has revealed that the five kingdom hierarchy may not be accurate. One kingdom, the Prokaryotae, can be divided into two major groups or domains: bacteria and archaea. The bacteria are fundamentally different from all other living things. The archaea are more similar to eukaryotes.

The relationship between classification and phylogeny

Classification and phylogeny

REVISED

Natural **classification** systems group living things according to their similarities. The more features that are shared between two organisms, the more closely related they are. **Phylogeny** is the evolutionary history or the evolutionary relationships between organisms and groups of organisms. Therefore, natural classification reveals the phylogeny. Phylogeny can be represented in an evolutionary tree such as Darwin's tree of life.

Now test yourself

TESTED

3 Explain why using DNA sequencing to classify organisms automatically matches their phylogeny.

Answer on p. 226

The evidence for the theory of evolution by natural selection

The contributions of Darwin and Wallace

REVISED

Charles Darwin made four observations in proposing his theory:

- Parents tend to produce more offspring than are able to survive.
- Populations tend to remain a constant size.
- Offspring look similar to their parents.
- No two individuals look identical.

From the above observations, Darwin devised the following theory:

- Parents produce too many offspring.
- There must be competition to survive.
- The better-adapted offspring survive, and pass on their features to the next generation.

Darwin called this theory 'evolution by **natural selection**'.

At the same time that Darwin was piecing his **theory of evolution** together, another scientist was doing the same thing. **Alfred Russel Wallace** collated evidence from parts of southeast Asia.

Natural selection is selection for certain features by natural forces.

Evidence for evolution

REVISED

Fossil evidence

Darwin used a lot of evidence from **fossils** to back up his theory of natural selection.

Fossils are found in sedimentary rocks and are formed when an organism leaves an imprint in soft mud or dies and comes to rest in the mud. As the mud hardens to form rock, the imprint or body remains in the rock.

Similarities between fossils can be used to reveal evolutionary relationships (phylogeny).

Molecular evidence

Certain chemicals, such as **DNA**, proteins and RNA, are universal to all living things. Variation is caused by changes in the DNA, which produce changes in proteins. As evolution occurs, the DNA accumulates more changes, as does the structure of the proteins it codes for. Therefore, closely related species have similar DNA and proteins, but more distantly related species have DNA and proteins that are more different.

The evidence from biochemistry is, perhaps, more reliable and more convincing than that from fossils.

Different types of variation

The term **variation** refers to the differences that exist between individuals.

Variation is the differences that arise between living organisms.

Intraspecific and interspecific variation

REVISED

Intraspecific variation occurs between members of the same species. These differences could be simple observable features such as colour, or they can be more obvious such as the differences between the sexes.

Variation can also occur between members of different species (**interspecific variation**). This depends on how closely related one species is to the other.

Exam tip

Remember that variation is the key to evolution — variation must occur before any characteristic can become beneficial and selected for.

Continuous and discontinuous variation

REVISED

Continuous variation is seen where there are no distinct groups or categories. There is a full range between two extremes. This form of variation is caused by:

- a number of genes interacting together
- the environment

Examples of continuous variation include height and body weight. The continuously variable feature can be quantified and data are usually presented in the form of a histogram.

Discontinuous variation is seen where there are distinct groups or categories. This type of variation is usually caused by one gene. The discontinuously variable feature cannot be quantified — it is qualitative — and data are usually presented in the form of a bar chart. Examples include gender and possession of resistance or immunity.

Continuous variation is variation that shows a complete range with no distinct groups.

Discontinuous variation is variation that produces distinct groups.

Revision activity

Write a list of features that are continuously variable and a separate list of features that are discontinuously variable.

Typical mistake

Students tend to plot bar charts with no spaces between the bars, but there should be gaps between them.

Causes of variation

REVISED

There are two causes of variation: genetic and environmental. Many variable features may be affected by both causes. For example, skin colour in humans is genetically determined. However, exposure to the sun results in the production of extra pigmentation, causing the skin to tan.

Genetic causes

Genetic causes of variation result from random mutations in the DNA sequence which are passed on from one generation to the next. Examples include: number of limbs, eye colour and ability to roll the tongue.

Environmental causes

Environmental causes of variation result from variations in exposure to certain environmental conditions. They are not passed from one generation to the next and cause continuous variation. Examples include:

- skin colour resulting from exposure to sunlight
- body mass

Now test yourself

4 Explain how genetic differences cause visible variation between members of the same species.

Answer on p. 226

TESTED

Revision activity

Write a list of features that show variation caused by genes, a separate list of features that show variation caused by the environment and a third list of features that show variation caused by both genes and the environment.

Adaptations of organisms

Adaptations enable the species to survive and thrive in their environments. These adaptations can be categorised as anatomical, physiological or behavioural.

Adaptations are features that help organisms to survive in their habitat.

Anatomical adaptations

REVISED

Anatomical adaptations are those that are associated with structure:

- Predators have sharp teeth to help kill and chew their prey. They also have a strong jaw joint so that it does not become dislocated by a struggling victim.
- Herbivores have a long and complex digestive system. This allows them to digest plant tissues.
- Plants have long, deep roots with many root hairs. This enables them to absorb water and minerals from the soil.
- Some plants such as the black mangrove, which lives in waterlogged soil, grow roots up into the air to gain oxygen above the anaerobic soil.

Physiological adaptations

REVISED

Physiological adaptations are those that are associated with how the body systems function:

- The kidneys of mammals extract water from the urine before excreting nitrogenous waste. This helps to reduce the need to find and drink water.
- Some plants, called C4 plants, collect carbon dioxide at night. They can keep their stomata closed during the hottest part of the day, reducing loss of water via transpiration.
- Yeasts respire anaerobically when there is no oxygen in their habitat. This means they can produce ATP and continue to grow.

Behavioural adaptations

REVISED

Behavioural adaptations are those that are associated with feeding, nesting and mating:

- Robins nest in a hole in a tree stump or wall a few inches above the ground. This means they are not competing with other bird species.
- In dry conditions, some plants open their stomata to make the leaves wilt. This reduces the surface area exposed to hot sun and reduces the rate of transpiration.

Convergent evolution

REVISED

In some cases, organisms from different taxonomic groups become adapted to the same habitat by adopting **similar anatomical features**. For example, the **marsupial mole** and the **placental mole** look remarkably similar although they are not closely related. They have independently evolved similar traits as a result of having to adapt to similar environments or ecological niches.

Typical mistake

Students often forget to explain how each adaptation helps survival of the species.

Exam tip

In examination questions, you are likely to be given some information about a particular species and its habitat. The question will then ask you to explain how the features described help the organism to survive.

How natural selection can affect the characteristics of a population

The mechanism of natural selection

REVISED

Genetic variation exists between individuals and some factor in the environment applies a **selection pressure**. Some variations are better adapted to survive than others, passing on their alleles to the next generation in greater numbers. Over a number of generations, the proportion of the population possessing these **advantageous characteristics** increases.

Now test yourself

TESTED

5 Draw a flow diagram to explain how a change in the environment can cause a change in the proportion of individuals possessing a particular characteristic.

Answer on p. 226

The implications of evolution

Pesticide resistance in insects

REVISED

When pesticides are used, they kill all susceptible **insects**. However, some individuals may have a degree of **pesticide resistance**. These few individuals may survive and breed to pass on their resistance to following generations. The pesticide has acted as a selective agent. As successive generations show some variation, it is possible for the insects to become increasingly resistant to higher and higher concentrations of the pesticide. The implications are that insects can damage food crops and, with no effective pesticides, we will be unable to prevent this damage.

Drug resistance in microorganisms

REVISED

Most bacteria are susceptible to antibiotics, but some **microorganisms** may show **drug resistance**. Just as with insect resistance to pesticides, the bacterial species eventually evolves resistance. The implications are that there are only a certain number of antibiotics and, once bacteria have evolved resistance to them all, we will have no further defences to help us combat disease.

Now test yourself

TESTED

6 Explain how strains of the bacterium *Clostridium difficile* can become resistant to antibiotics.

Answer on p. 226

Exam practice

1 The leopard *Panthera pardus* is a member of the cat family. Complete the following table to show its full classification. [5]

Kingdom	
	Chordata
Class	Mammalia
	Carnivora
Family	Felidae
Genus	
	pardus

2 (a) (i) State two features of continuous variation. [2]
(ii) State two features of discontinuous variation. [2]
(b) List the causes of variation. [2]
(c) Explain what is meant by the term *selection*. [3]

3 (a) Explain how inappropriate use of antibiotics has given rise to the so-called superbug methicillin-resistant *Staphylococcus aureus* (MRSA). [5]
(b) Explain how the structure of proteins can be used as evidence for evolution. [4]

Answers and quick quiz 13 online

ONLINE

Exam tip

Questions sometimes take the form of a description of the mechanism of evolution. This is likely to be in the context of artificial selection of farm animals or plants.

Exam tip

Remember that for new strains or new species to arise, variation and selection must occur over a number of generations.

Revision activity

Write a list of all the key terms used in this chapter, then add the meaning of each key term.

Summary

By the end of this chapter you should be able to:

- Describe the classification of species into the taxonomic hierarchy.
- Outline the characteristic features of the five kingdoms.
- Outline the use of the binomial system for nomenclature.
- Describe the evidence used in the classification of organisms.
- Discuss the fossil and biochemical evidence for evolution.
- Discuss variation within and between species.
- Describe the differences between continuous and discontinuous variation.
- Explain both genetic and environmental causes of variation.
- Outline the adaptations of organisms to their environments.
- Outline the roles of variation, adaptation and selection in evolution.
- Discuss the evolution of resistance to pesticides in insects and resistance to drugs in microorganisms.

14 Communication and homeostasis

It is essential that both plants and animals can respond to changing **stimuli** in their environment. **Responses** rely on communication within the body, which can be by electrical or chemical signals. **Homeostasis** is the control of the internal environment and relies on this communication. It controls conditions such as temperature, blood sugar concentration and blood water potential.

A **stimulus** is a change in the environment that causes a **response**.

Homeostasis is the maintenance of a constant internal environment.

The need for communication

Multicellular organisms must maintain a steady internal environment for the cells to function correctly. Any change in the external environment can cause internal changes. This means that any change in the environment must be monitored so that a suitable response is brought about to maintain the correct internal environment.

Communication between cells is essential to carry out a response that is suitable to the stimulus. A response may involve coordinating several organs and tissues. For example, running away from a predator involves the nervous system, skeleton and muscles, lungs and circulatory system as well as increased respiration within cells.

Revision activity

Write a list of all the external conditions that can act as stimuli to bring about a response.

Now test yourself

TESTED

1 For each internal parameter that is monitored write a note explaining why it must be kept constant. This explanation must include a detailed statement about what happens if the parameter gets too low or too high.

Answer on p. 226

Cell signalling

- Cells communicate by a process called **cell signalling**.
- The neuronal system and the hormonal system are communication systems that use cell signalling.
- The cells of the neuronal system are arranged close together so that the chemical signal does not need to travel far between cells. This makes the communication more rapid.
- The hormonal system uses the circulatory system to transport signalling molecules called hormones. As the signalling molecules can travel long distances, this form of communication is slower.

Cell signalling is communication between cells. It involves the release of a signalling molecule, which binds to the recipient cell and causes a change within the recipient cell.

Exam tip

There are many examples of cell signalling in living things, so you cannot be expected to know them all. You must be prepared to read the information given in the question and use it to construct your answer.

The principles of homeostasis

Receptors and effectors

REVISED

Responding to stimuli requires a complex sequence of events:

stimulus ⟶ receptor → communication pathway (cell signalling) ⟶ effector → response

input **output**

- The stimulus must be detected by a **receptor**.
- A receptor is a cell, tissue or organ that is sensitive to a particular form of stimulus and can detect changes in the level of that energy.
- Receptors are found on the surface of the body to detect changes in the external environment — for example, temperature, touch and pressure receptors in the skin.
- There are also receptors inside the body to detect changes in the internal environment — for example, temperature, water potential and pH sensors in the brain.
- Receptors convert changes in the environment (stimuli) to electrical signals that can stimulate the neuronal system.
- The neuronal system carries messages to the coordinating centre (part of the communication pathway); this is known as input.
- An **effector** is a cell, tissue or organ that brings about a response. These are often the muscles that enable movement or cells such as liver cells that can absorb or release glucose into the blood.
- The effectors are controlled by the coordinating centre, which sends messages to the effectors; this is known as output.

A **receptor** is a cell, tissue or organ that can detect changes in the environment.

An **effector** is a cell, tissue or organ that brings about a response.

Feedback

REVISED

The effectors respond to the output from the coordination centre. They change the conditions inside the body. This change is detected by the internal receptors, which modify the input to the coordination centre. This is known as feedback.

Negative feedback

Systems in the body monitor certain parameters such as internal temperature. These parameters need to be kept at a set point that corresponds to the optimum conditions. If the internal temperature changes away from the set point then a mechanism is put in place to reverse the change. This ensures that the internal temperature does not change too much and remains fairly constant. This is **negative feedback** (Figure 14.1).

Negative feedback is the response to a change in conditions that acts to reverse that change.

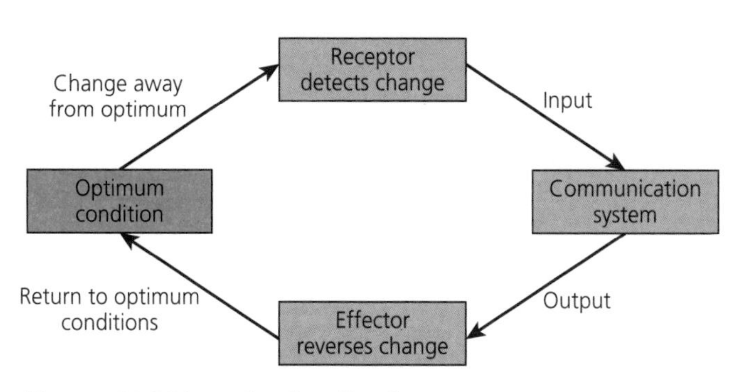

Figure 14.1 Negative feedback

Positive feedback

If a parameter such as temperature changes it might bring about further increases in that change. For example, as bacteria respire they produce heat and the increased temperature makes the bacteria more active, so they respire more and release even more heat.

Positive feedback is usually harmful, but there are one or two examples where it can be beneficial. In nervous conduction a small change in membrane potential opens ion channels that allow movement of ions through the membrane, which further increases the change in membrane potential to produce an action potential.

Positive feedback is when a small change brings about an increase in that change.

Now test yourself

2 Explain why positive feedback can only be used for a short period of time in living systems.

Answer on p. 227

TESTED

Maintaining body temperature

Maintaining body temperature is known as thermoregulation.

Ectotherms

REVISED

Ectotherms are organisms that rely on their surroundings to gain body heat. Their body temperature will always be dependent upon the surrounding environmental temperature. However, many ectotherms are able to regulate their body temperature to some extent by using behavioural adaptations:

- Lizards bask in the Sun in order to warm up. They hide in the shade or burrow when too hot.
- Many insects flap their wings to generate some heat in their wing muscles before flying.

Ectotherms are organisms that rely on external sources of heat.

Typical mistake

Many students refer to ectotherms as 'cold-blooded' — this is not the case, as some can keep their body temperature well above 30°C.

Revision activity

Draw a sketch outline of a reptile and add arrows to indicate each source of heat gain and loss. Annotate the arrows to explain how the heat is gained or lost.

Endotherms

REVISED

Endotherms use both behavioural and physiological mechanisms to regulate their body temperature.

Receptors in the hypothalamus of the brain monitor internal core temperature. The thermoregulatory centre in the hypothalamus uses this information to send out instructions to make adjustments that will reverse any change in core body temperature. Peripheral receptors in the skin detect changes in the temperature of the environment that also act as inputs to the thermoregulatory centre.

Endotherms can generate their own heat to regulate their body temperature more effectively.

Table 14.1 Physiological responses to change in temperature

Effector	Response if too hot	Response if too cold
Skin sweat glands	More sweat produced — evaporation cools skin	Less sweat produced
Skin hairs	Hairs lie flat	Hairs erected to trap air as insulation
Blood vessels in skin	Vasodilation — blood flows close to skin surface to lose heat by radiation and convection	Vasoconstriction — blood diverted to flow further from surface resulting in less radiation and convection
Muscles	Muscles relax	Muscles contract spasmodically to create heat — this is shivering
Liver	Reduces metabolic rate	Increases metabolic rate to release heat

Revision activity

Draw a sketch outline of a mammal and add arrows to indicate each source of heat gain and loss. Annotate the arrows to explain how the heat is gained or lost.

Now test yourself

TESTED

3 Explain why the skin is the main organ of thermoregulation.

Answer on p. 227

Revision activity

Write a list of all the key terms used in this chapter. Write the meaning of the key term next to each one.

Typical mistake

Students often refer to blood vessels in the skin moving closer to the surface or deeper down. The blood vessels do not move; they simply dilate or constrict to divert the blood flow.

Exam practice

1 (a) Explain the term *cell signalling*. [2]
(b) Name *two* systems in the mammalian body that use cell signalling. [2]
(c) Define the term *homeostasis*. [2]
(d) Explain why negative feedback is essential in a homeostatic mechanism. [4]

2 (a) In the early morning bees can be seen buzzing their wings but not flying. Explain the reasons behind this behaviour. [3]
(b) Explain how blood can be diverted to and from the skin and how this can be used to help thermoregulation in a mammal. [8]
(c) The graph shows the rate of flow of heat through the skin of a person during exercise.

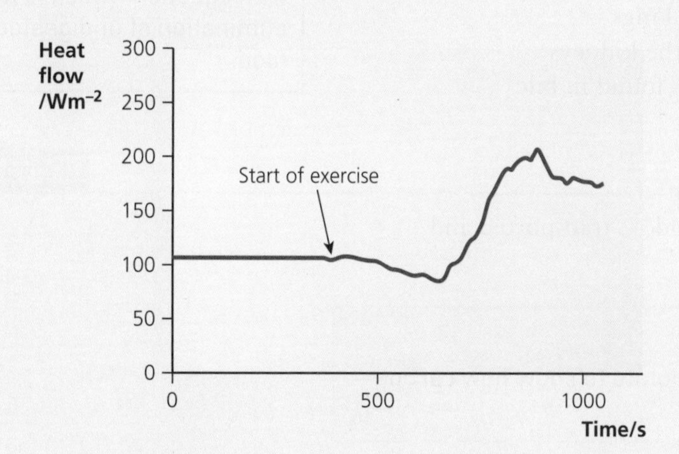

(i) Describe the changes shown in the graph. [3]
(ii) Explain the changes shown in the graph. [4]

Answers and quick quiz 14 online

ONLINE

Summary

By the end of this chapter you should be able to:

- outline the need for communication systems within multicellular organisms
- state that cells need to communicate with each other by a process called cell signalling
- define the terms negative feedback, positive feedback and homeostasis
- explain the principles of homeostasis
- describe the responses that maintain core body temperature in ectotherms and endotherms

15 Excretion as an example of homeostatic control

Excretion is the removal of metabolic waste from the body and is therefore part of homeostasis. The lungs, kidneys and liver are all involved in the removal of toxic metabolic products from the body. The kidneys also play a major role in controlling the water potential of the blood.

The liver also metabolises certain toxins that have been absorbed from the digestive system.

Excretion is the removal of metabolic waste from the body.

Exam tip

There are many topics that were learned earlier in the course which are important here. These include: the structure and ventilation of the lungs, the circulatory system, the structure of amino acids and the effect that changes in pH can have on protein structure and enzyme action.

Excretion

Metabolic wastes are the waste products from processes that have occurred inside the cells. They include:

- carbon dioxide, which is removed via the lungs
- nitrogenous waste, which is removed via the kidneys
- other substances, such as the bile pigments found in bile

Typical mistake

Do not confuse excretion with egestion, which is the elimination of undigested food.

Removal of carbon dioxide

REVISED

See chapter 8 for details of how carbon dioxide is transported and excreted.

Revision activity

Draw a diagram of a red blood cell and annotate to show how carbon dioxide is transported.

Removal of nitrogenous waste

REVISED

Nitrogenous waste includes ammonia, urea and uric acid. The liver cells, or hepatocytes, convert excess amino acids to urea which is then transported in the blood to the kidneys. The kidneys remove the urea from the blood to form part of the urine.

Now test yourself

TESTED

1 Explain why molecules such as carbon dioxide and ammonia should not be allowed to build up to a high concentration inside cells.

Answer on p. 227

The role of the liver

The liver consists of cells called hepatocytes. These cells are all the same and are metabolically very active. They contain a wide range of enzymes and organelles to help process substances in the blood. Some of the processed substances are excreted but others are stored for later use.

Now test yourself

TESTED

2 Describe how liver cells are adapted to their role.

Answer on p. 227

The gross structure of the liver

REVISED

The structure of the liver has evolved to deliver as much blood as possible to the hepatocytes. The liver receives blood from two sources:

- The hepatic artery carries in oxygenated blood from the aorta.
- The hepatic portal vein carries in deoxygenated blood from the digestive system.

Inside the liver these vessels divide up to form narrow branches that flow along the portal areas between the many liver lobules. Each lobule consists of many hepatocytes arranged in columns radiating out from the centre of the lobule, like the spokes of a wheel (Figure 15.1). The blood from both sources mixes as it enters a lobule and flows along a narrow channel called a sinusoid. This channel runs between the hepatocytes towards the centre of the lobule.

As blood flows past the hepatocytes exchange occurs across the cell surface membranes. Many substances are removed from the blood to be processed in the hepatocytes, while other substances such as glucose may be released into the blood.

At the centre of each lobule the sinusoids flow into a branch of the hepatic vein. The hepatic vein carries blood back out of the liver (Figure 15.2).

(a)

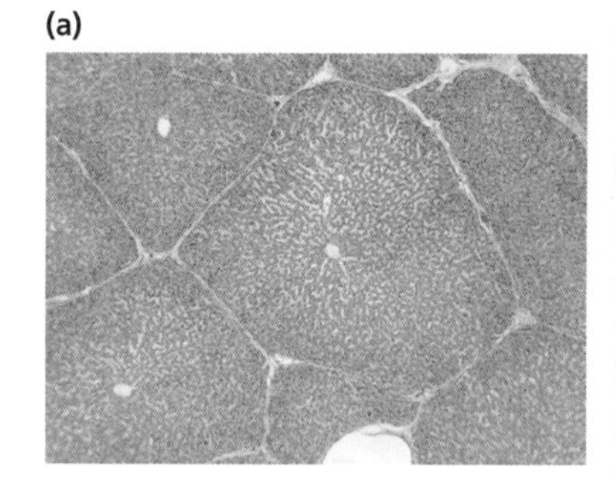

(b)

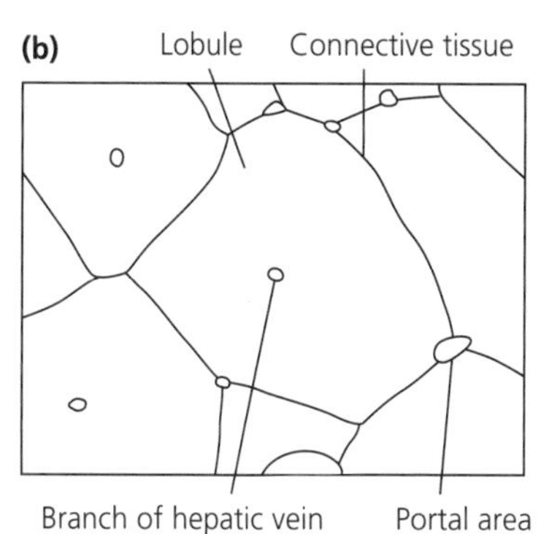

Figure 15.1 (a) Low-power photograph of liver lobule. (b) A plan drawing of this photograph

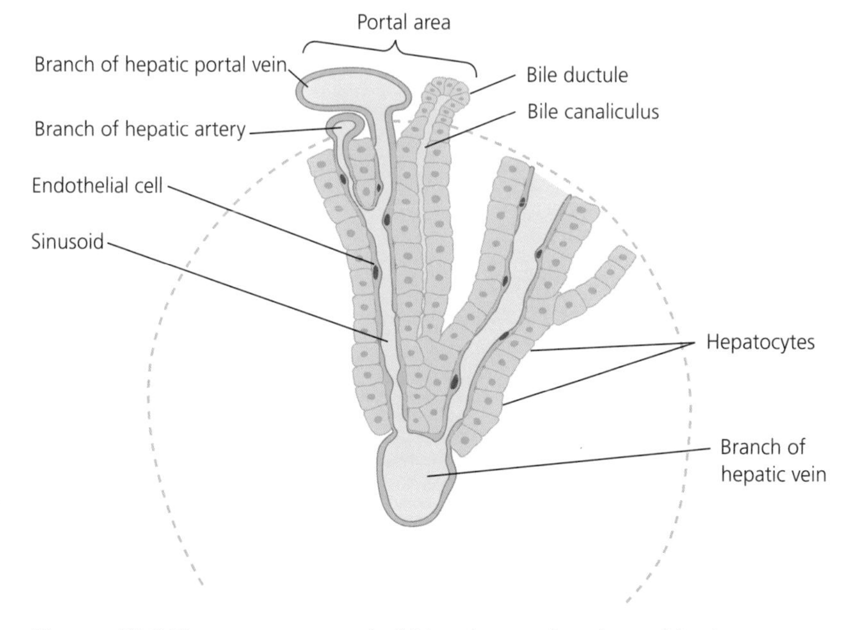

Figure 15.2 The arrangement of blood vessels, sinusoids, hepatocytes and bile canaliculi inside a liver lobule

Bile canaliculi also run between the columns of hepatocytes. These carry bile in the opposite direction from the sinusoids. The bile carries certain substances towards the bile duct, which lies in the portal area between adjacent lobules. These substances include:

- bile salts, which are used in digestion
- bile pigments, which are excretory products

Typical mistake

Many students confuse the intralobular vessel (the hepatic vein) with the interlobular vessels in the portal area (the hepatic artery and the hepatic portal vein).

Now test yourself

TESTED

3 Explain why the liver receives blood from two sources.

Answer on p. 227

Formation of urea

REVISED

Excess amino acids are **deaminated** in the hepatocytes. This involves the removal of the amino group to form ammonia. It leaves an organic acid residue that can be used in respiration:

amino acid → organic acid residue + ammonia

Deamination is the removal of the amino group from an amino acid.

The ammonia is highly toxic and must quickly be converted to the less toxic nitrogenous compound urea:

ammonia + carbon dioxide → urea + water

This is achieved in the ornithine cycle (Figure 15.3).

Exam tip

Study the ornithine cycle carefully — some questions do ask for quite a bit of detail.

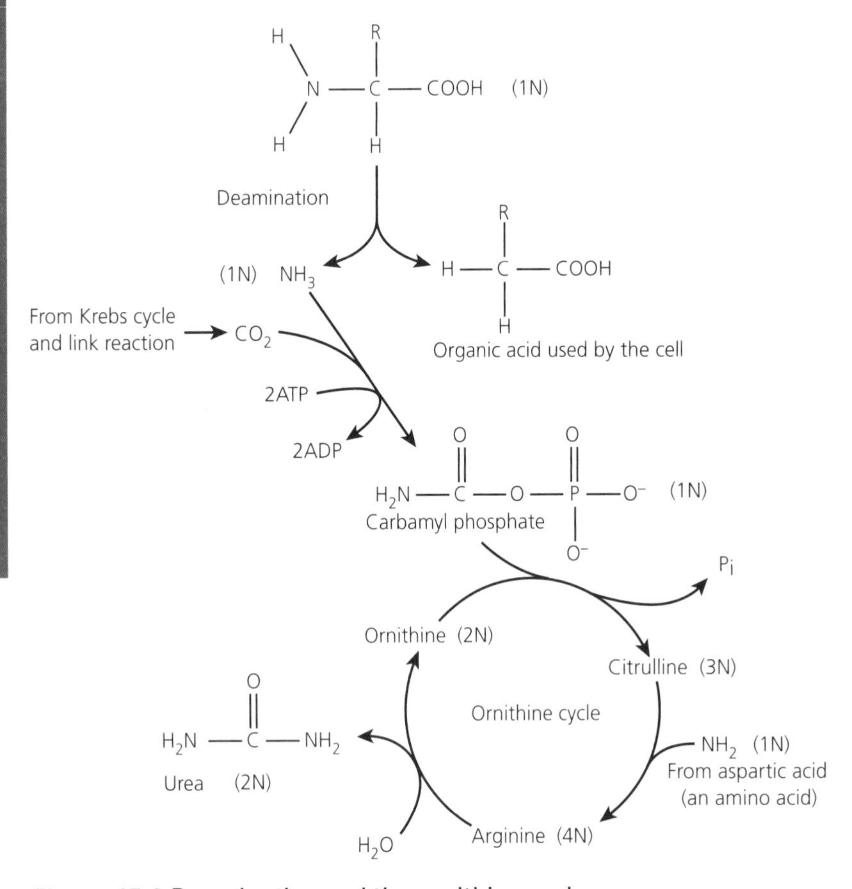

Figure 15.3 Deamination and the ornithine cycle

Now test yourself

TESTED

4 Explain why mammals must convert ammonia to urea while fish can excrete ammonia directly into their surroundings.

Answer on p. 227

Detoxification

REVISED

The liver has a number of other roles in metabolism. These include:

- storing carbohydrates in the form of glycogen
- detoxification — in particular the detoxification of hydrogen peroxide, alcohol (which is oxidised to ethanal and then to ethanoic acid before being converted to acetyl coenzyme A) and other drugs such as paracetamol, steroids and antibiotics

Revision activity

Sketch a liver cell and annotate with arrows to show the movement of substances into and out of the cell.

The kidney

Structure of the kidney

The kidney is supplied with blood from the renal artery. Inside the kidney the blood passes into specialised capillary beds (glomeruli), which filter the blood before it returns to the body via the renal vein. Waste products in the blood enter the ureter to flow to the bladder. The kidney consists of two layers — the outer cortex and inner medulla. In the centre is the pelvis, which collects the urine produced and leads into the ureter (Figure 15.4).

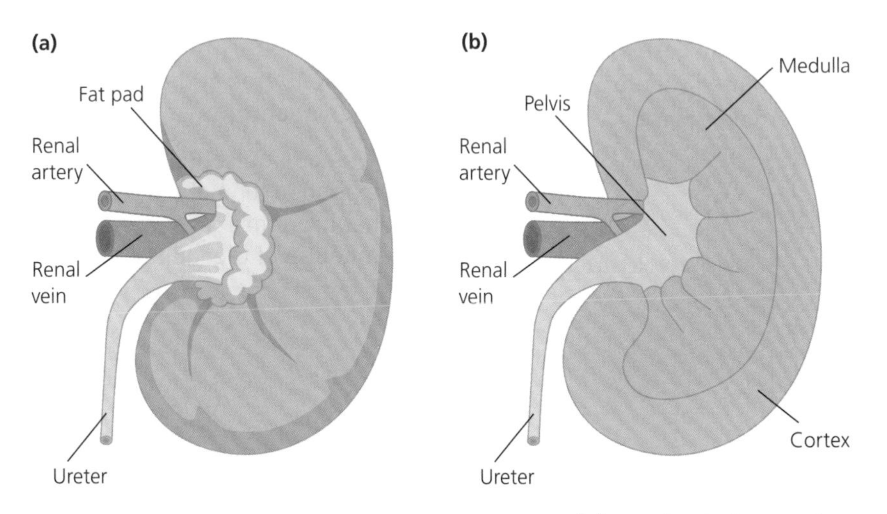

Figure 15.4 The gross structure of the kidney: (a) external view and (b) vertical section

Each kidney contains around one million tiny tubules called nephrons (Figure 15.5). These start in the cortex at the Bowman's capsule, which surrounds each glomerulus. Fluid filtered out of the blood enters the capsule and follows a convoluted path along the nephron into the collecting ducts that lead into the pelvis.

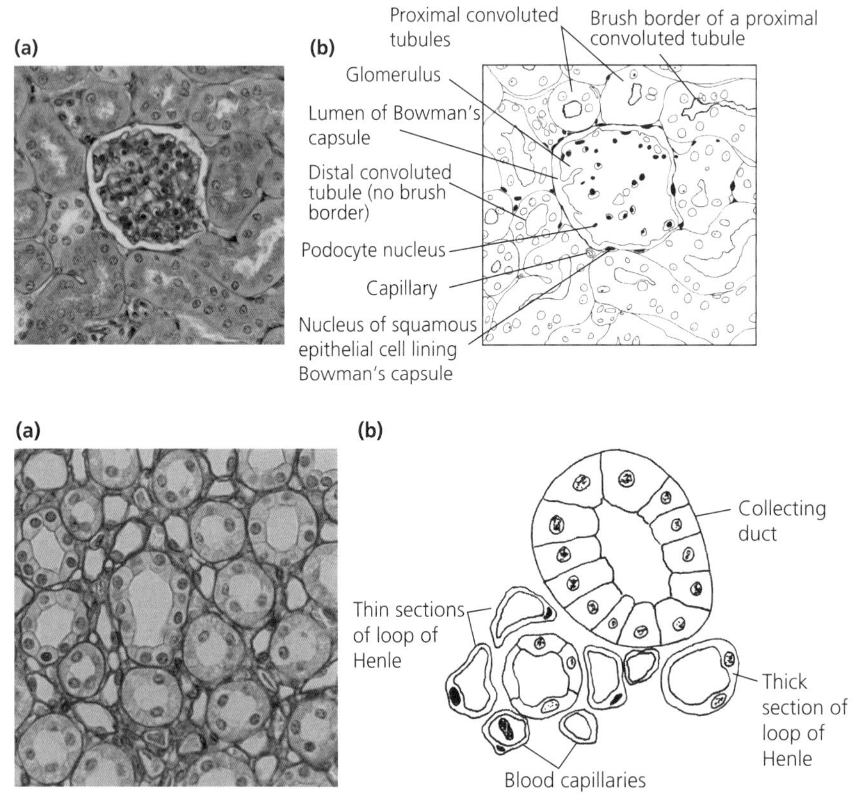

Figure 15.5 (a) Photomicrograph of a glomerulus (× 300) and (b) a diagram of the micrograph. (c) Photomicrograph of tubules in the medulla (× 4000) and (d) a diagram of the micrograph.

The nephron

REVISED

Each **nephron** consists of a Bowman's capsule, which surrounds the **glomerulus** (Figure 15.6). This is where ultrafiltration takes place. Fluid passes from the capsule to the proximal convoluted tubule, where selective reabsorption takes place. The fluid then passes along the **loop of Henle**, down into the medulla and back out to the cortex. From the loop of Henle the fluid passes into the distal convoluted tubule and into the collecting duct, where reabsorption of water takes place.

A **nephron** is a tiny kidney tubule that filters the blood and produces urine.

A **glomerulus** is a tiny knot of capillaries at the start of each nephron.

The **loop of Henle** is an extended loop of the nephron that runs into the medulla of the kidney.

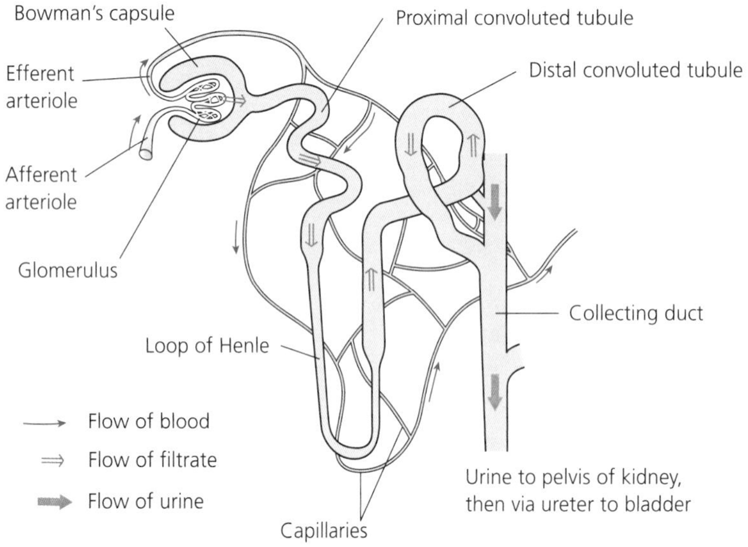

Figure 15.6 A kidney nephron and associated blood vessels

Bowman's capsule
Proximal convoluted tubule
Efferent arteriole
Distal convoluted tubule
Afferent arteriole
Glomerulus
Collecting duct
Loop of Henle
→ Flow of blood
⇒ Flow of filtrate
➡ Flow of urine
Urine to pelvis of kidney, then via ureter to bladder
Capillaries

Action of the nephron

REVISED

Ultrafiltration

- The arteriole leading into the glomerulus (afferent arteriole) is wider than the arteriole leading out (efferent arteriole). This maintains the blood pressure in the capillaries.
- Blood plasma containing the dissolved components of blood is squeezed out of the capillaries.
- The Bowman's capsule surrounding the glomerulus has a specialised inner layer of cells called **podocytes**. These have major (primary) processes and minor (secondary) processes, which leave small gaps between the cells that allow the fluid to enter the capsule.
- Between the capillary of the glomerulus and the podocytes is a basement membrane — this is the filter and it allows the passage of molecules with a relative molecular mass of 69 000 or less (Figure 15.7).

Ultrafiltration is the filtering of the blood through a basement membrane, which acts as a very fine sieve.

Podocytes are specialised cells lining the Bowman's capsule that leave gaps between them to reduce the resistance to flow of fluid.

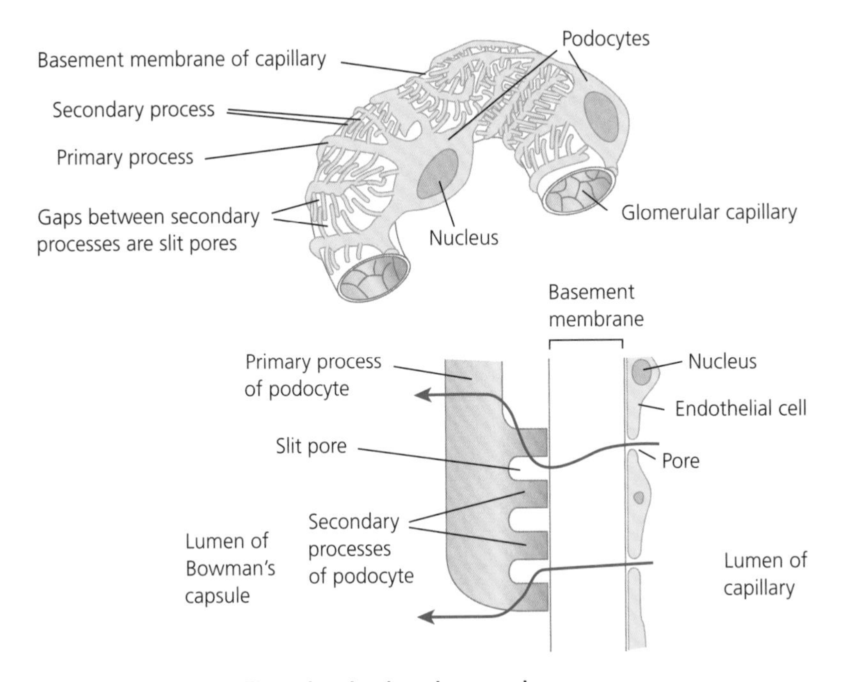

Figure 15.7 Ultrafiltration in the glomerulus

Selective reabsorption

The cells of the proximal convoluted tubule are specialised to reabsorb glucose, amino acids and mineral ions from the fluid in the tubule. The membranes of these cells contain sodium–potassium ion pumps and cotransport proteins. The membranes are folded into microvilli to increase their surface area. The cells also contain many mitochondria to supply ATP for active transport.

- Sodium ions are pumped out of the cell through the basal membrane towards the blood capillary. This is active transport and requires ATP.
- This reduces the concentration of sodium ions inside the cell, creating a concentration gradient.
- Sodium ions diffuse into the cell from the fluid in the tubule — but they diffuse through co-transport proteins that bring glucose molecules (or amino acids) into the cell at the same time as the sodium.
- The concentration of glucose in the cell rises and it can diffuse out through the basal membrane into the blood capillary (Figure 15.8).

Selective reabsorption is the removal of certain substances from the fluid in the tubule, leaving other substances in the tubule.

Typical mistake

Many students confuse 'active process' with 'active transport'. The reabsorption of glucose is an active process, which includes the active transport of sodium ions out of the cells lining the tubule.

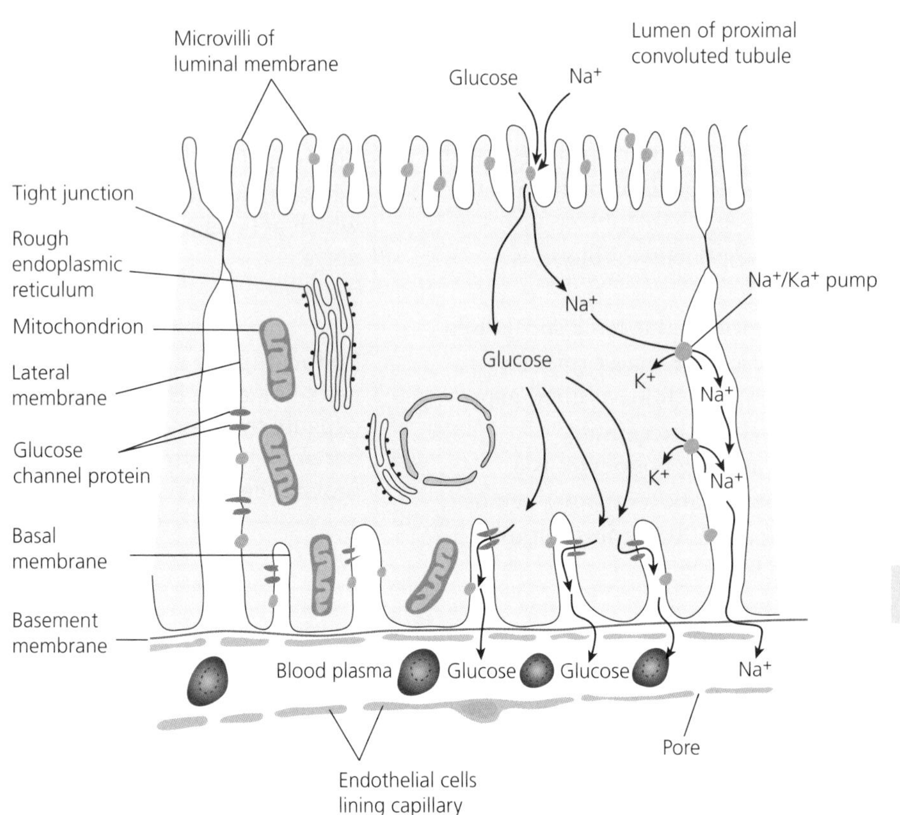

Figure 15.8 Selective reabsorption by cells in the proximal convoluted tubule

Now test yourself

5 Explain why it is not essential to reabsorb proteins from the fluid in the nephron.

Answer on p. 227

TESTED

The role of the loop of Henle

The loop of Henlé modifies the fluid in the tubule to help conserve water.

- As the fluid passes down the loop it becomes more concentrated due to salts diffusing into the fluid from the surrounding tissue.
- As the fluid moves up the loop it becomes more dilute as salts are pumped out of the fluid in the tubule.
- This passes salts from the ascending limb to the descending limb so that the fluid becomes increasingly concentrated. It is called a **hairpin countercurrent multiplier**.
- The effect is to create a concentration gradient in the medulla of the kidney. The salt becomes more and more concentrated towards the centre of the kidney.
- This means that the water potential of the tissue fluid is very low.
- As fluid passes along the collecting duct from the cortex to the pelvis water is withdrawn by osmosis into the tissue fluid.
- This results in urine that has a lower water potential than blood and conserves water.

A **hairpin countercurrent multiplier** is a mechanism to increase the concentration of substances in a tube where substances are passed between two parts of the same tube in which the fluid flows in opposite directions.

Exam tip

The way in which the loop of Henle works is complicated — remember that its role is to produce a salt concentration gradient in the medulla. This is another useful area in which the examiner can test your knowledge of AS material and osmosis.

Typical mistake

Many students get confused with water potentials and concentrations — never use the term water concentration and remember that a higher salt concentration causes a lower water potential.

The control of water potential in the blood

Osmoreceptors in the hypothalamus monitor the water potential of the blood. When blood water potential falls too low it causes water to be withdrawn from the osmoreceptor cells by **osmosis** and the cells shrink. Shrinkage causes the osmoreceptors to stimulate neurosecretory cells that lead down into the posterior pituitary gland.

The neurosecretory cells release antidiuretic hormone (ADH) directly into the blood. The ADH travels around the body but its target tissue is the collecting ducts in the kidney. It binds to cell surface receptors on the cells lining the collecting ducts and causes the release of cAMP inside the cell as the second messenger. The end result is the insertion of more water permeable channels into the membrane lining the lumen of the collecting duct (Figure 15.9). This makes the cells more permeable to water. More water is reabsorbed from the urine in the collecting duct and the blood water potential rises again.

If the blood water potential rises too high the osmoreceptor cells swell and less ADH is released.

Now test yourself

TESTED

6 Explain why a mechanism such as this is called negative feedback.

Answer on p. 227

Exam tip

This is an excellent example of negative feedback.

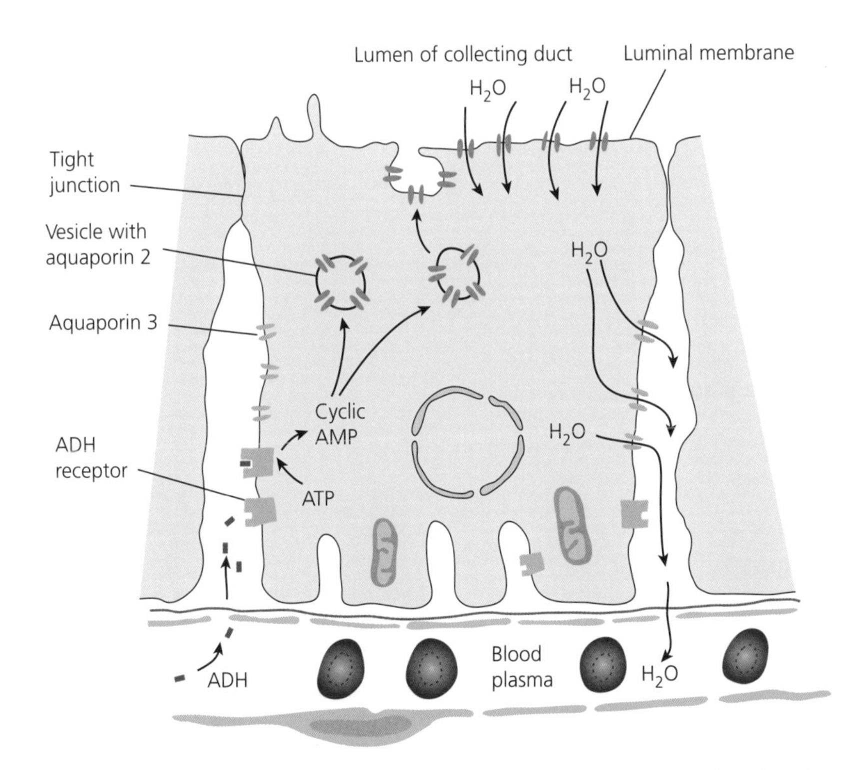

Figure 15.9 The changes that occur in a cell lining the collecting ducts due to release of ADH

Revision activity

Draw a flow diagram to show the responses to blood water potential getting too high and too low. Start with the set point as a horizontal line and show above the line an increase in water potential and the sequence of events to bring the water potential back to the set point. Repeat below the line for a drop in water potential.

Kidney failure

Kidney failure can be fatal because metabolic wastes such as urea, toxins and excess salt and water are not excreted. One of the most immediate effects is an imbalance in electrolytes in the blood and body fluids. Kidney failure is detected by changes in glomerular filtration rate (GFR). GFR measures how much fluid enters the nephrons each minute. A normal reading is in the range 90–120 $cm^3 min^{-1}$. If the GFR drops below 60 $cm^3 min^{-1}$ it indicates some form of chronic kidney disease. A reading below 15 $cm^3 min^{-1}$ indicates kidney failure.

Kidney failure can be treated in two ways:

- Renal dialysis involves connection to a dialysis machine, which takes blood from a vein in the arm and passes it over a dialysing membrane. The membrane is partially permeable and exchanges materials between the blood and special dialysis fluid. The fluid contains all the correct concentrations of substances in the blood and must be constantly refreshed. Over time the unwanted substances in the blood, such as urea, diffuse into the dialysing fluid and are removed. The blood is returned to a vein in the arm. Peritoneal dialysis is an alternative in which dialysing fluid is placed in the abdominal cavity for exchange to occur across the peritoneum. The disadvantage of dialysis is that the patient may need to spend many hours a week connected to the machine and also needs to monitor dietary intake carefully.
- Kidneys can be transplanted successfully and the patient can then live a normal life. However, care must be taken to ensure a good match between the donor and recipient.

Revision activity

Draw a mind map to link together all the possible consequences of kidney failure.

Urine tests

Drugs and hormones are treated by the liver and will be excreted by the kidneys in urine. Urine can be tested for the presence of the hormones, drugs or traces of drugs. Use of recreational drugs such as cannabis can leave traces in the urine for several days.

Pregnancy tests

REVISED

Human embryos release a hormone called human chorionic gonadotrophin (hCG), which appears in the urine of the mother. Pregnancy testing sticks contain monoclonal antibodies specific to this hormone. The sticks contain antibodies that are attached to tiny blue beads. These antibodies are free to move. Other antibodies are attached (immobilised) in a line across the stick. If the hormone is present it binds to the free antibodies and to the immobilised antibodies, holding the beads in a line across the stick.

Revision activity

Write a list of all the key terms used in this chapter. Write the meaning of the key term next to each one.

Anabolic steroid tests

REVISED

Some athletes are tempted to use anabolic steroids to boost their muscle growth and strength. Urine can be tested by gas chromatography, in which a gaseous solvent is used to separate substances in the urine, or by mass spectrometry. Use of anabolic steroids gives an unfair advantage and has been banned in sport.

Exam practice

1 (a) List *three* functions of hepatocytes. [3]
(b) Describe the arrangement of hepatocytes in the liver and explain how this ensures that as much blood as possible can be treated by the hepatocytes. [4]

2 (a) Name the region of the nephron where selective reabsorption takes place. [1]
(b) Describe how selective reabsorption of amino acids is achieved. In your answer you should use appropriate technical terms, spelt correctly. [5]
(c) Name one other substance that is selectively reabsorbed. [1]
(d) Describe *three* features of the cells of this region of the nephron that adapt them to their role of selective reabsorption and explain how each adaptation helps the process. [6]

3 The table shows the ratio of concentrations of substances found in the capillaries relative to the concentration found in the Bowman's capsule.

Substance	Concentration in capillary relative to Bowman's capsule
Protein	15 000:1
Amino acid	1:1
Glucose	1:1
Urea	1:1
Sodium ions	1:1

(a) Name the process that occurs between the capillary and the Bowman's capsule. [1]
(b) Identify one type of molecule that cannot enter the Bowman's capsule. [1]
(c) Describe what prevents this substance entering the Bowman's capsule. [2]
(d) Explain why the concentration of amino acids in the Bowman's capsule is the same as that in the capillary. [2]

Answers and quick quiz 15 online

ONLINE

Summary

By the end of this chapter you should be able to:

- define the term excretion and explain the importance of removing metabolic wastes from the body
- describe the histology and gross structure of the liver
- describe the formation of urea in the liver, including the ornithine cycle
- describe the roles of the liver in detoxification
- describe the histology and gross structure of the kidney
- describe the detailed structure of a nephron and its associated blood vessels
- describe and explain the production of urine and explain the control of blood water content
- outline the problems arising from kidney failure and how kidney failure can be treated
- describe how urine samples can be tested for pregnancy and misuse of anabolic steroids

16 Nerves

Sensory receptors

Sensory receptors monitor conditions in the surroundings or inside the body. They only respond when the conditions change. Such a change in conditions is called a stimulus. Receptors are transducers. They convert the stimulus into electrical energy in the form of an action potential. There is a wide range of sensory receptors. Each type of receptor can detect a specific stimulus. For example, the temperature receptors detect changes in temperature and the Pacinian corpuscles detect changing pressure on the skin.

A **sensory receptor** is a specialised cell that converts one form of stimulus into an action potential.

Revision activity

Write a list of the different sensory receptors and the types of energy that they respond to.

Neurones

Neurones are the nerve cells. There are three types (Table 16.1).

A **neurone** is a specialised cell that can transmit an action potential

Table 16.1

Feature	Sensory neurones	Motor neurones	Relay neurones
Function	Transmit action potentials from the sensory receptors to the central nervous system	Transmit action potentials from the central nervous system to the effectors, such as muscles	Transmit action potentials between the sensory and motor neurones; found in the central nervous system
Myelin sheath	Both possess a myelin sheath to speed up conduction and insulate the neurone from surrounding cells		Non-myelinated
Cell body	Positioned just outside the central nervous system in a swelling called the dorsal root ganglion	Positioned inside the central nervous system	Positioned inside the central nervous system
Dendrites	Many	Many	Variable — few to many
Dendron	One long dendron	No dendron	No dendron
Axon	One short axon	One long axon	Many have short axons; some have a long axon
Ending	Synaptic knob	Motor end plate	Synaptic knob
Diagram	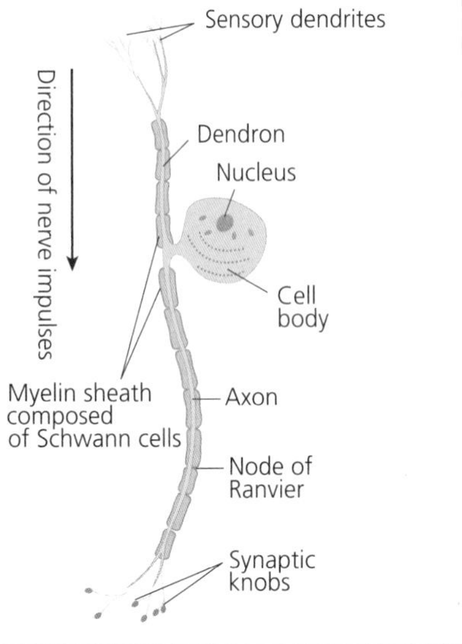 Sensory dendrites Direction of nerve impulses Dendron Nucleus Cell body Myelin sheath composed of Schwann cells Axon Node of Ranvier Synaptic knobs	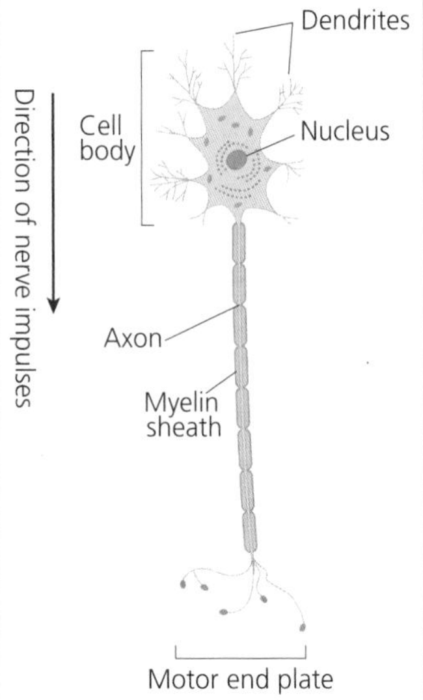 Dendrites Cell body Nucleus Direction of nerve impulses Axon Myelin sheath Motor end plate	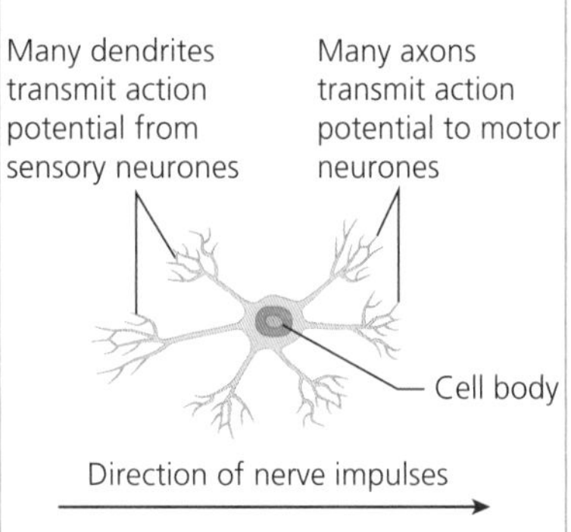 Many dendrites transmit action potential from sensory neurones Many axons transmit action potential to motor neurones Cell body Direction of nerve impulses

Now test yourself

TESTED

1 Explain why a motor neurone has a long axon while a sensory neurone has a short axon.

Answer on p. 227

Revision activity

Sketch a diagram of the three types of neurone. Annotate the diagram to show how these cells are adapted to their function.

Resting potentials

When a neurone is at rest it maintains a potential difference across its membrane — the membrane is polarised. This is called the **resting potential**. The inside of the cell is maintained at −60 to −70 mV compared with outside the cell. The neurone membrane contains sodium–potassium pumps, which use ATP. They pump three sodium ions out of the cell for two potassium ions into the cell. Therefore, while at rest, the cell contains many potassium ions and few sodium ions.

A **resting potential** is the potential difference across the cell membrane when the cell is at rest.

Action potentials

An **action potential** is a depolarisation of the membrane caused by an influx of sodium ions. The action potential can move along the membrane as a nerve impulse.

An **action potential** is the change in membrane potential that is transmitted along the neurone.

Generation of an action potential

REVISED

When a neurone is stimulated it becomes depolarised — this means it loses its normal polarisation. The neurone membrane contains special voltage-gated sodium ion channels that open as a result of a change in the potential difference. These allow sodium ions to flow by facilitated diffusion into the cell. As a result, the inside of the cell becomes less negative compared with outside. If enough sodium ions flow into the cell the potential across the membrane reaches the threshold potential, which opens more sodium channels — this is positive feedback. More sodium ions flow into the neurone and the potential inside the cell rises to about +40 mV compared with outside. This is an action potential (Figure 16.1).

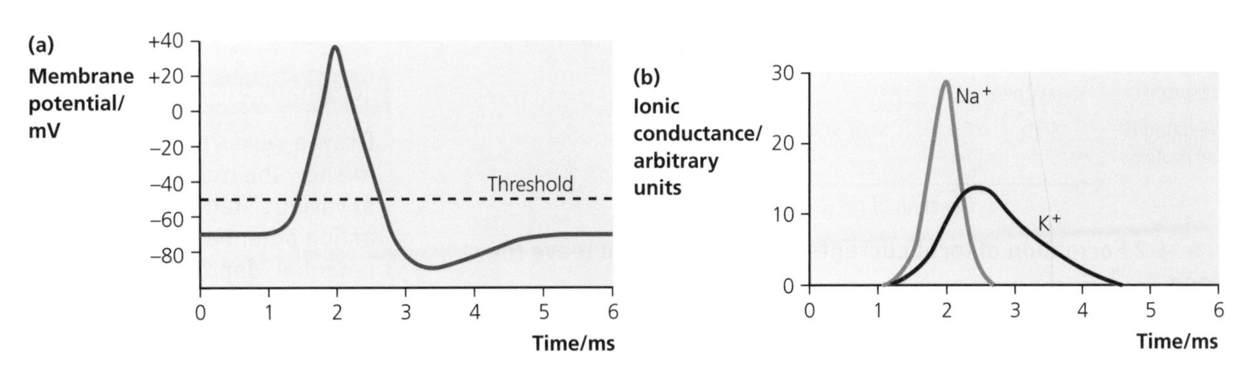

Figure 16.1 The change in (a) potential difference and (b) conductance to sodium and potassium ions during an action potential

After the sodium ions have entered the cell other gated ion channels open to allow potassium ions out of the cell. This reduces the potential difference across the membrane again, returning it to the −60 mV resting potential. This is called repolarisation. However, the membrane potential does briefly fall below the normal resting potential. This is called hyperpolarisation.

Now test yourself

TESTED

2 Explain why the first part of an action potential is called depolarisation and the second part is called repolarisation.

Answer on p. 227

Typical mistake

Many students are not sure how the ions move across the membrane during an action potential. The resting potential maintains a concentration gradient across the membrane — so sodium ions diffuse across the membrane into the cell and potassium ions diffuse out. After the action potential these ions are pumped by active transport back to their original positions.

Exam tip

The creation of action potentials and their transmission along the neurone is a complex process. Break it down into small steps and remember the sequence of events.

Transmission of an action potential

REVISED

As sodium ions enter the neurone they diffuse along the inside of the neurone. This produces a local current, which alters the potential difference across the membrane and causes sodium ion channels further along the neurone to open (Figure 16.2). This allows more sodium ions to enter, and so on, and the action potential moves along the neurone.

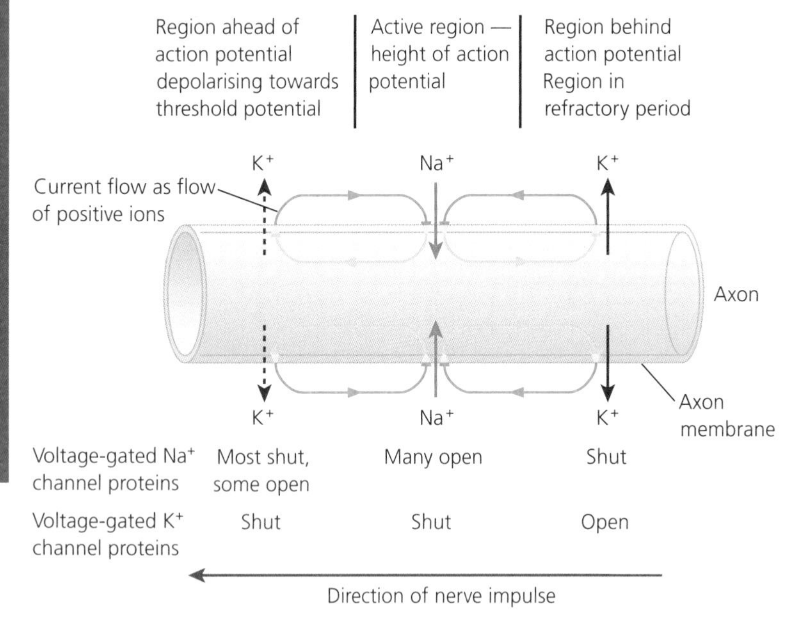

Figure 16.2 Formation of local currents as ions enter and leave the neurone

After the potassium ions diffuse into the neurone the membrane is back to its resting potential. However, because the ions are effectively in the wrong places, the sodium-potassium pumps have to re-establish the correct positioning of the ions. Until this happens the axon enters a refractory period in which a new impulse cannot be generated.

Revision activity

Draw a series of diagrams to show the membrane at various stages of the action potential (resting potential, depolarisation and repolarisation). The diagrams should show the activity of the sodium ion channels, the potassium ion channels and the sodium–potassium pumps.

Myelinated and unmyelinated neurones

Many neurones are myelinated. This means that each individual neurone is surrounded by a fatty **myelin sheath** created by individual Schwann cells wrapped around the neurone. Between the Schwann cells are gaps called **nodes of Ranvier**, which occur at 1 mm intervals along the neurone. Ions cannot move across the membrane where the myelin sheath is in place — only at the nodes. In these neurones the local currents are stretched, to carry the depolarisation from one node to the next. This speeds up transmission of the impulse because the action potential jumps between nodes in a process called saltatory conduction.

Unmyelinated neurones are still surrounded by Schwann cells (Figure 16.3). However, one Schwann cell surrounds several neurones and there are no nodes of Ranvier. In unmyelinated neurones the action potential does not jump along the neurone.

A **myelin sheath** is a non-conducting fatty layer around the neurone.

A **node of Ranvier** is a gap in the myelin sheath.

Now test yourself

3 Explain how the action potential gets from one node of Ranvier to the next.

Answer on p. 227

TESTED

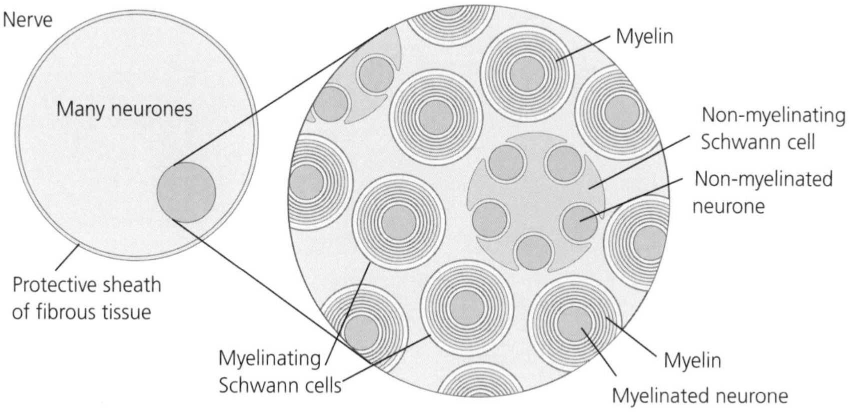

Figure 16.3 Myelinated and unmyelinated neurones

Nervous communication

All neurones conduct action potentials in the same way. Once the threshold potential has been reached all actions potentials are identical — this is known as the all-or-nothing rule. Therefore there is no difference in the signal received by the brain from different stimuli. So how does the brain analyse the incoming information and interpret the stimulus so that we understand what the stimulus is and how intense it is?

The neurones are connected into particular pathways. Neurones from one particular type of sensory receptor run to the same area of the brain. Therefore, when this part of the brain is stimulated it is interpreted as that particular stimulus.

The intensity of the stimulus will affect the frequency at which the neurones fire — a higher intensity will create more frequent impulses in the neurones.

The structure of a synapse

Neurones do not actually touch one another where they meet at a **synapse**. There is a small gap between them called a synaptic cleft. Therefore an action potential cannot pass from one neurone to another directly.

A **synapse** is a junction between two neurones.

The gap is bridged by the release of a **neurotransmitter**.

Neurones are well adapted to cell signalling:

- The first neurone ends in a bulge called the presynaptic knob.
- This provides space to store vesicles of neurotransmitter.
- It also provides a large surface area for the release of neurotransmitter molecules (acetylcholine).
- The second neurone contains specialised proteins in its membrane.
  - They act as acetylcholine receptor sites.
  - They form sodium ion channels, which open in response to acetylcholine molecules.

A **neurotransmitter** is a chemical that diffuses across the gap or cleft between two neurones. Acetylcholine is the neurotransmitter in cholinergic synapses.

Transmission across the synapse

Transmission across a synapse involves the release of acetylcholine into the synaptic cleft and the detection of that acetylcholine on the postsynaptic membrane (cell signalling):

1. An action potential travelling along the presynaptic neurone.
2. Causes calcium ion channels to open so calcium ions diffuse into the knob.
3. Vesicles of acetylcholine move towards membrane.
4. Vesicles fuse with presynaptic membrane to release acetylcholine molecules into the cleft (exocytosis).
5. The acetylcholine molecules diffuse across the cleft and bind to the acetylcholine receptor sites on the postsynaptic membrane.
6. Sodium ion channels in the postsynaptic membrane open.
7. Sodium ions diffuse into the postsynaptic neurone.
8. Depolarisation of the membrane.
9. The enzyme acetylcholinesterase breaks down the acetylcholine.
10. Sodium ion channels close and the choline is recycled back into the presynaptic knob (Figure 16.4).

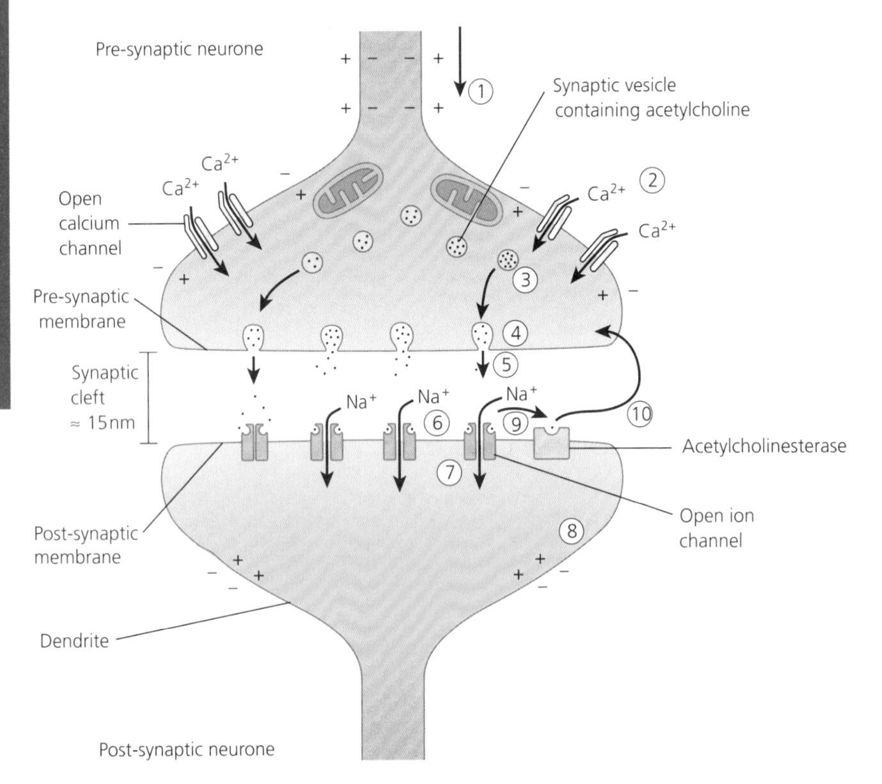

Figure 16.4 The events occurring during transmission at a synapse

Typical mistake

Many students write a description that makes it sound as if the vesicles are released and diffuse across the cleft. The vesicles fuse to the membrane and release the acetylcholine molecules.

Revision activity

Draw a diagram of a synapse and annotate it to show how transmission occurs across the synapse.

The role of synapses in the nervous system

Synapses join neurones together, allowing transmission of a signal from one neurone to another (cell signalling). Action potentials travelling in the wrong direction can be stopped as there are no vesicles of acetylcholine in the postsynaptic neurone. Impulses from low-intensity stimuli can be filtered out so that there is no unnecessary response. This is achieved because many vesicles must be released to cause a postsynaptic action potential. Continuous, unimportant stimuli can be ignored as the vesicles of acetylcholine run out after a while. This called fatigue and prevents continuous responses to continuous, unimportant stimuli. This allows a form of behaviour called acclimatisation.

Action potentials in one neurone can be used to stimulate several postsynaptic neurones so that multiple responses can be achieved from one stimulus. Nerve impulses from more than one stimulus can be combined to create the same response — this can be used to magnify or amplify the response to a low-intensity stimulus. This is called summation.

Inhibitory synapses can prevent formation of an action potential in the postsynaptic neurone.

Revision activity

Write a list of all the key terms used in this chapter. Write the meaning of the key term next to each one.

Exam practice

1 (a) (i) Identify the molecules that prevent movement of charged particles across a membrane. [1]
(ii) Describe how the membrane of a motor neurone is specialised to allow movement of charged particles. [3]
(b) Small ions such as Na^+ and K^+ leak across membranes. Explain how a neurone maintains the resting potential across its cell surface membrane despite this leakage. [3]
(c) Use the information in figure 16.1 to describe how an action potential is created. [5]

2 (a) (i) The presynaptic knob contains many organelles. Suggest three organelles that can be found in unusually high numbers. [3]
(ii) State the function of each of these organelles. [3]
(b) (i) Name the neurotransmitter used in cholinergic synapses. [1]
(ii) Describe the mechanism by which the neurotransmitter is released. [4]
(c) Modern headphones can come with 'sound cancelling technology'. These can work by emitting a constant frequency behind the music being played. Suggest how this could work to cancel other background sounds. [2]

3 Complete the following paragraph.
A synapse is a junction between two .... The action potential in the presynaptic neurone causes the release of ...... from vesicles in the presynaptic knob. The neurotransmitter molecules ...... across the synaptic cleft and bind to ...... molecules in the postsynaptic membrane. These molecules have a specific shape that is ...... to the shape of the neurotransmitter molecule. The binding of the neurotransmitter molecules opens ...... ion channels and ions flow ...... the postsynaptic neurone. This ...... the postsynaptic membrane, producing a new .... [9]

Answers and quick quiz 16 online

ONLINE

Summary

By the end of this chapter you should be able to:

- outline the roles of sensory receptors in mammals
- describe the structures and functions of sensory, relay and motor neurones
- describe and explain how an action potential is generated and transmitted in a neurone
- describe the structure of a cholinergic synapse and the role of neurotransmitters in the transmission of action potentials

17 Hormonal communication

The endocrine system

The endocrine system is a group of glands that release hormones directly into the blood. These glands are not linked other than by the blood circulatory system and the hormones that are transported in the blood.

Endocrine glands

REVISED

Endocrine glands are glands that have no ducts. They secrete chemical signalling molecules (hormones) directly into the blood.

An **endocrine gland** is a ductless gland that produces hormones and secretes them directly into the blood.

Typical mistake

Students often forget to mention that the hormones are released directly into the blood.

Hormones

REVISED

Hormones are complex molecules that act as chemical signals. They are released from an endocrine gland directly into the blood and are transported in the blood to their target organ or target tissue. The molecule has a specific shape. The site at which the hormone acts is called the target cell or target tissue. The cells of the target tissue have cell surface receptors that have a complementary shape to the shape of the hormone.

A **hormone** is a cell-signalling molecule that is transported in the blood.

Now test yourself

TESTED

1 Explain why hormones are transmitted all over the body in the blood and yet affect only the correct target tissues.

Answer on p. 228

First and second messengers

REVISED

First messenger

The hormone is known as the **first messenger**. It carries the signal from the endocrine gland to the cells of the target tissue.

The **first messenger** is the hormone; the **second messenger** is the molecule that transmits the signal inside the target cell.

Second messenger

The **second messenger** is a molecule inside the cells of the target tissue that transmits the signal from the membrane-bound receptor into the cell cytoplasm, where an effect can be carried out. One example of a first and second messenger is the effect that adrenaline has on its target cells:

- Adrenaline in the blood binds to the membrane-bound receptor (an adrenergic receptor).
- A G-protein is activated that, in turn, activates the enzyme adenyl cyclase.

- The adenyl cyclase converts adensosine triphosphate (ATP) to cyclic adenosine monophosphate (cAMP).
- The cAMP acts a second messenger (Figure 17.1) by moving into the cell cytoplasm and causing an effect in the cell — in this case by activating specific enzymes.

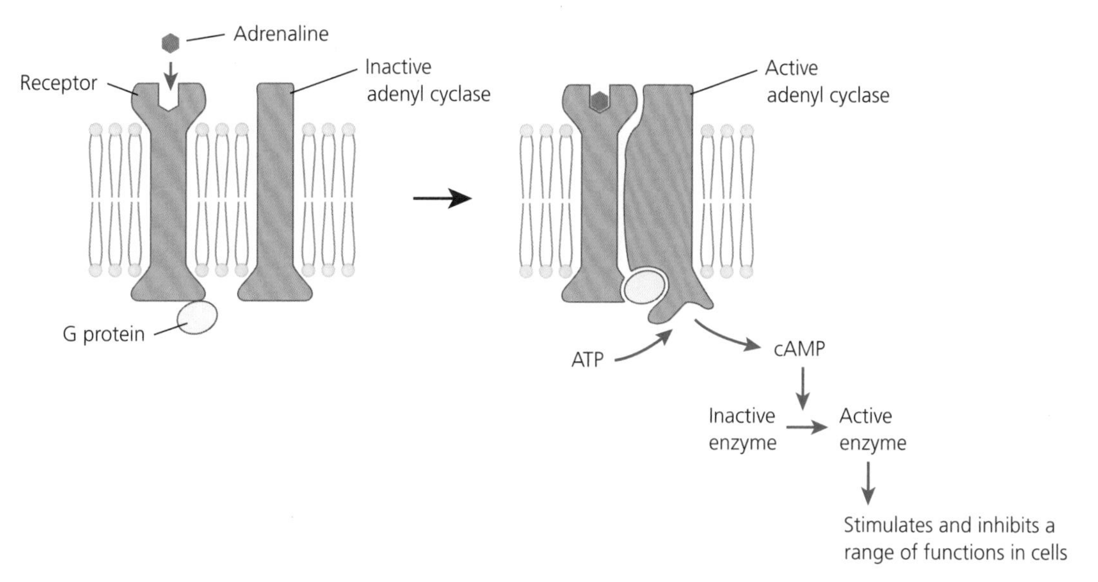

Figure 17.1 The activation of cAMP as a second messenger

Now test yourself

TESTED

2 Suggest how cAMP can have different effects in different cells.

Answer on p. 228

The adrenal glands

The adrenal glands are endocrine glands that lie just above the kidneys. They have two regions — an outer cortex and an inner medulla.

The cortex releases steroid-based hormones called adrenocorticoids. These contribute to the homeostatic maintenance of glucose and mineral concentrations in the blood, which also affect blood pressure. For example, cortisol stimulates production of glucose from stored compounds; aldosterone acts on the distal tubules in the kidney to increase absorption of sodium ions.

The adrenal medulla releases adrenaline, which prepares the body for activity. This includes the following effects:

- stimulating the breakdown of glycogen
- increasing blood glucose concentration
- increasing heart rate
- increasing blood flow to muscles
- decreasing blood flow to gut and skin
- increasing the width of bronchioles, to ease breathing
- increasing blood pressure

Typical mistake

Students often forget that the adrenal glands have two regions, and omit to describe the adrenal cortex and its hormones.

Now test yourself

3 Write a brief explanation of how each effect of adrenaline listed helps to prepare the body for activity.

Answer on p. 228

TESTED

The pancreas

The pancreas (Figure 17.2) is both an exocrine gland and an endocrine gland.

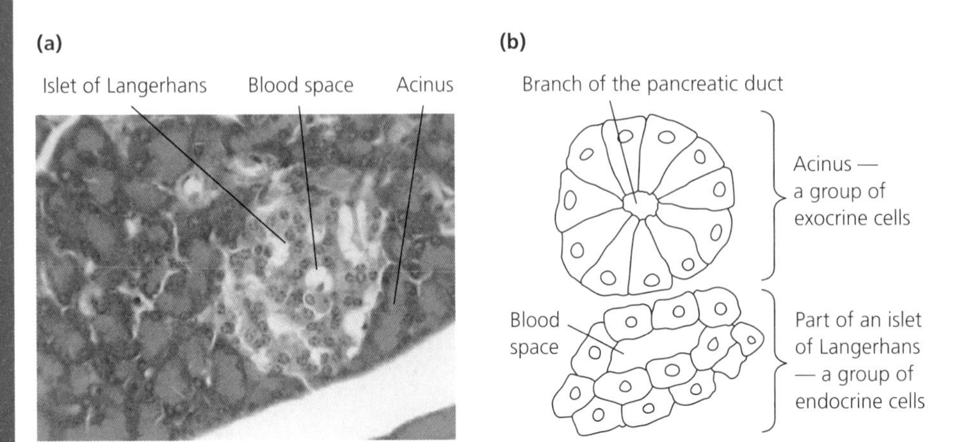

Figure 17.2 (a) A high-power photomicrograph of the pancreas. (b) A drawing showing cells from the exocrine and endocrine regions

Exocrine function

REVISED

The majority of the pancreas consists of enzyme-producing cells arranged into acini (singular acinus). An acinus is a group of cells arranged around a tiny ductule. The cells manufacture digestive enzymes that are released into the ductule. Many ductules combine to form the pancreatic duct. The pancreatic duct carries the enzymes into the small intestine.

Endocrine function

REVISED

In between the exocrine tissue are patches of endocrine tissue. These are called **islets of Langerhans**. These consist of two types of cell: the α cells manufacture and release glucagon directly into the blood. Glucagon has the effect of increasing blood glucose concentration. The β cells manufacture and release insulin. Insulin decreases blood glucose concentration.

The **islets of Langerhans** are small patches of endocrine tissue in the pancreas.

Exam tip

In your written responses you must be specific — refer to 'blood glucose concentration' rather than 'sugar concentration' or 'level of glucose'. Also, make sure that all terms such as 'glycogen', 'glucagon' and 'glucose' are spelt correctly.

Regulation of blood glucose concentration

Glucose concentration in the blood is maintained at about 90 mg 100 cm^{-3} or 4–5 mmol dm^{-3}. This is the set point. The concentration varies depending on what is eaten and the level of activity.

If blood glucose concentration gets too high

REVISED

- High glucose concentration.
- Stimulates release of insulin into the blood. Insulin binds to receptors on the membranes of liver cells.
- The liver cells absorb more glucose and convert it to glycogen; they use more glucose in respiration and inhibit conversion of fats and glycogen to glucose.
- Insulin also causes muscle cells to absorb more glucose.

If blood glucose concentration gets too low

REVISED

- Low blood glucose stimulates the α cells to release glucagon.
- Glucagon binds to specific membrane-bound receptors.
- These activate adenyl cyclase to produce cAMP, which acts as a second messenger inside the liver cells.
- Glucagon causes the liver cells to increase the conversion of glycogen to glucose and stimulates increased conversion of fats and amino acids to glucose.

Typical mistake

Many students do not learn the names of the substances well enough and often give spellings that suggest they are hedging their bets, for example 'glucagen' or 'glycogon'.

Revision activity

Draw a flow diagram to show the responses to blood glucose concentration getting too high and too low. Start with the set point as a horizontal line and show above the line an increase in glucose concentration and the sequence of events to bring the concentration back to the set point. Repeat below the line for a drop in concentration.

Controlling the release of insulin

REVISED

1 At normal concentrations of glucose the β cells maintain a membrane potential of −60 to −70 mV inside through the action of sodium/potassium ion pumps. Open potassium ion channels allow diffusion of potassium ions out of the cell.
2 High blood glucose concentrations cause glucose molecules to diffuse into the β cells.
3 The glucose is used in respiration to manufacture ATP.
4 Increasing ATP concentration closes the potassium channels which stops potassium ions leaving the cell.
5 The concentration of potassium ions inside the cell increases causing partial depolarisation of the membrane to −30 mV.
6 Calcium ion channels open and calcium ions diffuse into the β cells.
7 Vesicles containing insulin move to, and fuse with, the plasma membrane.
8 Insulin is released from the cell by exocytosis (Figure 17.3).

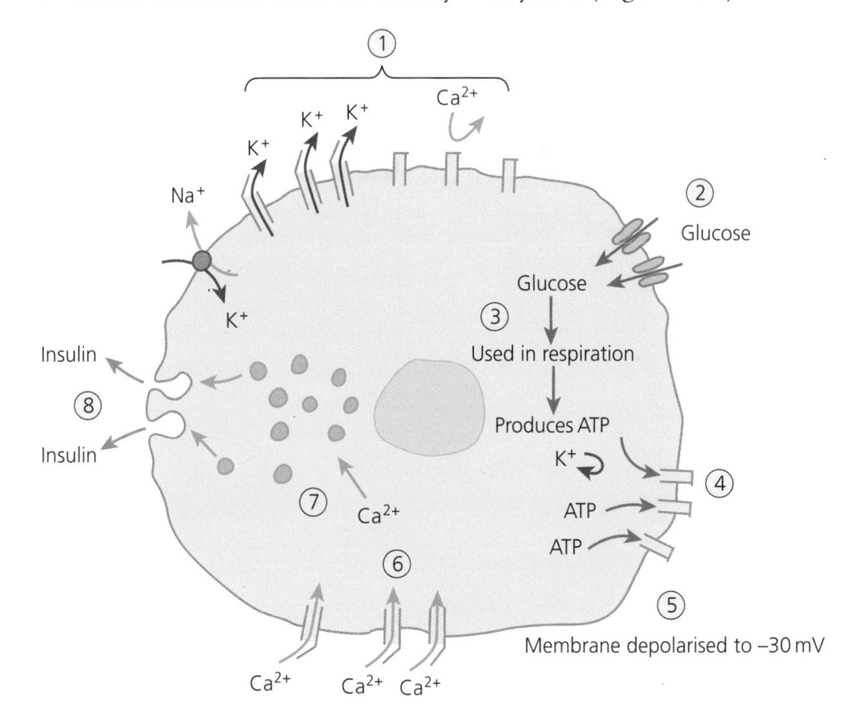

Figure 17.3 The release of insulin from beta cells

Revision activity

Convert Figure 17.3 into a flow chart describing the sequence of steps in the release of insulin.

Now test yourself

4 Draw a table to compare the release of insulin from β cells to the action of a synapse.

Answer on p. 228

TESTED

Exam tip

Remember that answers can be written in the form of bullet points or numbered points — this will help you to keep track of how many points you have made.

Diabetes

Diabetes mellitus is a treatable condition in which the body is unable to control the concentration of glucose in the blood. There are two types of diabetes mellitus (Table 17.1).

Diabetes mellitus is the inability to control blood glucose concentration.

Table 17.1

Feature	Type 1 diabetes	Type 2 diabetes
Uncontrolled blood glucose	Yes	Yes
Insulin dependent	Yes	No
Onset	Usually juvenile	Usually in middle age
Cause	Inactive β cells; possibly caused by an autoimmune response or a viral infection	Liver cells become less responsive to insulin Linked to obesity, high levels of refined sugars in the diet and to family history More common in people of Asian or Afro-Caribbean origin
Treated by insulin injections	Yes	No
Treated by careful monitoring of the diet and regular exercise	Yes	Yes

Now test yourself

TESTED

5 Explain why a person with diabetes mellitus might feel tired and be unable to exercise strenuously or for long periods.

Answer on p. 228

Potential treatments for diabetes

Insulin production by bacteria

REVISED

Insulin for injection to treat type 1 diabetes used to be isolated from the pancreases of pigs slaughtered for food. Genetically engineered bacteria can now be used to manufacture human insulin. This has a number of advantages:

- Human insulin is more effective than pig insulin.
- There is less chance of rejection.
- It is cheaper to produce.
- It is easier to increase production if demand increases.
- There are fewer ethical or moral objections.

Exam tip

The use of genetically modified bacteria to produce insulin is an example of how scientific advances benefit society — this is one of the aspects of How Science Works.

Use of stem cells

REVISED

Research is continuing into the possibility of using stem cells to grow new β cells in the pancreases of patients with type 1 diabetes. This would be a permanent cure and would free diabetes patients from the need to inject insulin every day.

Exam tip

The potential use of stem cells is another aspect of How Science Works. Questions are likely to test your ability to make use of the information you have been given.

Revision activity

Write a list of all the key terms used in this chapter. Write the meaning of the key term next to each one.

Exam practice

1 (a) Describe the regulation of blood glucose concentration. [5]
(b) Describe how the β cells in the pancreas respond to high blood glucose concentration. [3]
(c) Glucokinase is an enzyme that facilitates phosphorylation of glucose to glucose-6-phosphate in the cells of the pancreas. It plays an important role in the regulation of carbohydrate metabolism by acting as a glucose sensor, triggering shifts in metabolism or cell function in response to rising or falling levels of glucose. Mutations of the gene for this enzyme can cause unusual forms of diabetes. Suggest how a modified version of this enzyme could cause symptoms of diabetes. [2]

2 (a) Describe the difference between a first messenger and a second messenger. [2]
(b) Suggest why polypeptide hormones need to act through a receptor and a second messenger. [2]
(c) Explain how the pancreas can be both an endocrine gland and an exocrine gland. [3]

3 (a) Distinguish between type 1 and type 2 diabetes. [3]
(b) Suggest what advice might be given to a person with type 2 diabetes. [3]
(c) State the potential advantages of treating type 1 diabetes with stem cells. [2]

Answers and quick quiz 17 online

ONLINE

Summary

By the end of this chapter you should be able to:

- describe endocrine communication by hormones
- describe the structure and function of the adrenal glands
- describe the histology of the pancreas and outline its role as an exocrine and an endocrine organ
- explain how blood glucose concentration is regulated
- outline how insulin secretion from β cells is controlled
- compare and contrast the causes of type 1 and type 2 diabetes mellitus
- discuss the potential treatments for diabetes mellitus

18 Plant and animal responses

The types of plant response

All organisms need to respond to changes in their internal or external environment. The changes are stimuli and the responses made act to increase the survival chances of the organism.

Plants respond to avoid herbivory (being eaten) and to **abiotic** stress (insufficient light and water).

Abiotic factors are non-living factors.

Avoiding herbivory

REVISED

- Plants such as oak release tannins. These phenolic compounds are toxic to microorganisms and herbivores.
- Some plants also produce alkaloids, which make the tissue taste bitter and deter herbivory.
- Another response to herbivory is seen in the sensitive mimosa plant (*Mimosa pudica*). The leaves of this plant fold up in response to touch; this movement scares away insect herbivores.

Abiotic stress

REVISED

Plants respond to a shortage of water by reducing water loss. This can be achieved by closing their stomata. Such behaviour is coordinated by plant hormones, especially abscisic acid.

Responses to other abiotic stresses such as a shortage of light are achieved by growth movements called **tropisms**.

A **tropism** is a directional growth movement.

Tropisms

Tropisms involve directional growth movement:

- Phototropism — the response to light. Shoots show positive phototropism and grow towards light.
- Geotropism — the response to gravity. Shoots show negative geotropism and grow away from gravity, up into the light. Roots show positive geotropism and grow towards gravity, down into the soil.
- Chemotropism — the response to certain chemicals. The pollen tube grows towards the ovum during reproduction.
- Thigmotropism — the response to touch. Some plants grow so that their stem or a branch winds around a support.

Now test yourself

TESTED

1 Explain why a plant responding to touch might enhance its chances of survival.

Answer on p. 228

Plant hormones

Plant responses are coordinated by growth regulators, which are often called hormones.

Leaf loss (abscision) in deciduous plants

REVISED

Deciduous plants shed their leaves each year. There is a region of cells at the end of the leaf stalk called the abscission zone.

- Auxin produced in the leaf inhibits abscission by making the cells insensitive to ethylene.
- Cytokinins enter the leaf and stop the leaves aging.

When the concentration of cytokinins in the leaf drops the leaf ages. This is called senescence and stops the production of auxin in the leaf.

The cells in the abscission zone become sensitive to ethylene, which stimulates the production of the enzyme cellulase in the abscission zone. Cellulase digests the cell walls in the abscission zone and the leaf falls.

Seed germination

REVISED

When a seed takes up water this activates a hormone called gibberellin. The gibberellin travels to the aleurone layer, which surrounds the endosperm (starch store) of the seed. Gibberellin stimulates the production of amylase, an enzyme that hydrolyses stored starch to form glucose. The glucose provides a substrate for respiration and protein synthesis in the embryo so it can start to grow.

Stomatal closure

REVISED

Stomata can open and close through the action of the guard cells. Guard cells possess receptors for the hormone abscisic acid. Binding of abscisic acid activates a number of chemical pathways inside the cell, which lead to an increase in pH and a transfer of calcium ions from the vacuole to the cytoplasm. The calcium stimulates loss of charged ions (K^+, NO_3^- and Cl^-) from the cell. This increases the water potential of the cell so that water moves out of the cell by osmosis. The loss of turgor pressure changes the shape of the cell and the stoma closes.

Experimental evidence

Auxin and apical dominance

REVISED

Apical dominance refers to the fact that auxin is produced at the tip of the shoot (the apex) and that it inhibits the growth of lateral buds. The evidence for this includes:

- Removal of the apical bud allows the lateral buds to grow.
- Auxin or synthetic auxin placed on the cut tip continues to inhibit the growth of side shoots.

More recent investigations have revealed that the link is not direct. Auxin promotes production of abscisic acid, which inhibits the lateral bud growth.

Apical dominance refers to the inhibition of lateral shoots by the apical bud.

Gibberellins and stem elongation

REVISED

Elongation of the stem is stimulated by gibberellins. The evidence for this includes:

- Growing seedlings with certain fungi has been shown to increase seedling growth by causing elongation of the stem. Gibberellins can be isolated from these fungi.
- Applying a cream containing gibberellins extracted from fungi will also cause stem elongation.

Exam tip

In terms of factual recall the plant responses topic is quite light, however, you are expected to be able to answer questions based on experimental evidence.

Typical mistake

Many students fail to look at data provided in tables and graphs closely enough.

Gibberellins and seed dormancy

REVISED

Experiments have shown that early germination can be caused by application of gibberellins to a seed. Further evidence is seen in experiments that use hormone inhibitors. Germination is normally inhibited by abscisic acid. Without abscisic acid a seed will germinate in the fruit as it ripens. Inhibiting the gibberellins prevents this early germination.

Revision activity

Compile a table in which first column is the hormone and the second column is the effects on plant growth caused by that hormone.

Commercial uses of plant hormones

Table 18.1 outlines the commercial uses of plant hormones.

Table 18.1

Hormone	Commercial uses
Auxins	Promoting root growth in cuttings Producing seedless fruit As a selective weedkiller
Gibberellins	Delaying fruit senescence and drop to make harvesting more efficient Improving the shape and size of fruit Activating enzymes in stored barley to produce malt for brewing Speeding up seed production during breeding programmes
Cytokinins	Delaying leaf senescence to avoid discolouring of vegetables — this increases shelf life Promoting bud and shoot growth during tissue culture Promoting growth of lateral buds
Ethene	Speeding up fruit ripening and promoting fruit drop Promoting growth of lateral branches

Exam tip

You may be asked to make direct comparisons between the way plants respond and the way animals respond.

Revision activity

Write a list of all the key terms used in this chapter. Write the meaning of the key term next to each one.

Animal responses

Animals **respond** to:

- the threat of predation (being eaten)
- abiotic stress, which includes:
  - extremes of temperature
  - insufficient water
  - irritants
- internal **stimuli**, such as the need to find food and increase the chances of reproduction

A **stimulus** is a change in the environment that elicits a **response**.

A **response** is a change in behaviour resulting from a **stimulus**.

Revision activity

Draw a mind map with 'response' at the centre — include both plants and animals and note how and why they respond.

Typical mistake

Students often forget to consider the survival advantage of responses.

Organisation of the nervous system

The nervous system must detect environmental changes (stimuli) and bring about suitable responses to those changes. There are receptors all over the body to detect a variety of environmental stimuli — both internal and external. The nerves transmit the information and the brain receives information from all the receptors. The brain puts all the information together in association centres and coordinates suitable responses via the motor areas.

The nervous system can be divided into two regions:

- The central nervous system (CNS) consists of the brain and spinal cord.
- The peripheral nervous system (PNS) consists of all the receptors, the sensory neurones and the motor neurones. It can be divided into the somatic system (which deals with conscious movements and responses) and the autonomic nervous system.

The autonomic nervous system

REVISED

The **autonomic nervous system** is the part of the peripheral nervous system that coordinates unconscious responses to do with maintaining the internal environment.

The **autonomic nervous system** is a series of motor neurones that coordinate the unconscious responses involved in homeostasis.

All the cells are motor neurones and most are unmyelinated (Figure 18.1). The pathway carrying action potentials from the CNS to the effector consists of at least two neurones with synapses inside swellings called ganglia (singular = ganglion).

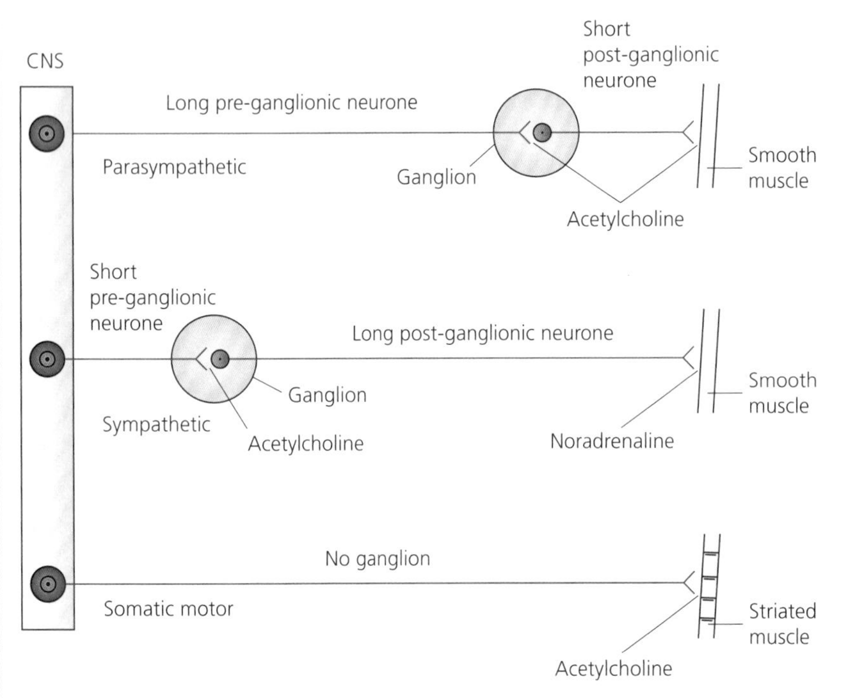

Figure 18.1 The neurones in the autonomic nervous system compared with those in the somatic system

The autonomic nervous system consists of two antagonistic sets of nerves: the parasympathetic nerves and the sympathetic nerves (Table 18.2).

Table 18.2

System	Parasympathetic system	Sympathetic system
Organisation	Only a few nerves leading out of the CNS (including the vagus nerve), which divide up and lead to different organs	Many nerves leading out of the CNS
Position of ganglia	In the effector tissue	Just outside the CNS
Length of pre-ganglionic fibres	Long	Short
Length of post-ganglionic fibres	Short	Long
Neurotransmitter	Acetylcholine	Noradrenalin
Role	Decreases activity — to conserve energy	Increases activity — to prepare for activity
When active	Most active in sleep or relaxation	Most active at times of stress
Effects	Decreases heart rate Constricts pupil Reduces ventilation rate	Increases heart rate Dilates pupils Increases ventilation rate

Now test yourself

TESTED

2 Explain why there are two parts to the autonomic nervous system.

Answer on p. 228

Revision activity

Draw an outline of a human body and inside the outline draw a simple version of the nervous system — include the autonomic system. Annotate your diagram to show the function of each part of the nervous system.

The structure and functions of the brain

The brain is divided into four sections each with its own function (Figure 18.2):

- Cerebrum — the largest part of the human brain, which is divided into two hemispheres joined by the corpus callosum. The cerebral cortex (the outer part of the cerebrum) is highly folded and deals with the 'higher functions' such as: conscious thought, overriding some reflexes, intelligence, reasoning, judgement and memory. The cerebral cortex is divided into:
  - sensory areas, which receive impulses from the sensory neurones
  - motor areas, which send impulses out to the effectors
  - association areas, which link information together and coordinate the appropriate response
- Cerebellum — coordinates fine control of muscular movement; for example, walking, running and using tools. It coordinates balance and body position and receives information from the various senses (retina, balance organs in the ear, muscles spindles, joints) and sends impulses to the motor areas.
- Medulla oblongata — controls the involuntary processes such as heart rate and breathing. It contains specialised centres such as the cardiac centre and the respiratory centre, which receive information from internal receptors and modify the heart rate and breathing rate accordingly.
- Hypothalamus and pituitary complex — controls the homeostatic mechanisms. The hypothalamus receives information from a variety of receptors — mostly internal — monitoring the blood and regulating factors such as body temperature and blood water potential. It also controls the endocrine system via the pituitary gland, which is known as the master gland. It releases a variety of hormones, which control the activity of certain organs and other endocrine glands.

Revision activity

Draw a sketch of the brain and annotate with the functions of each part.

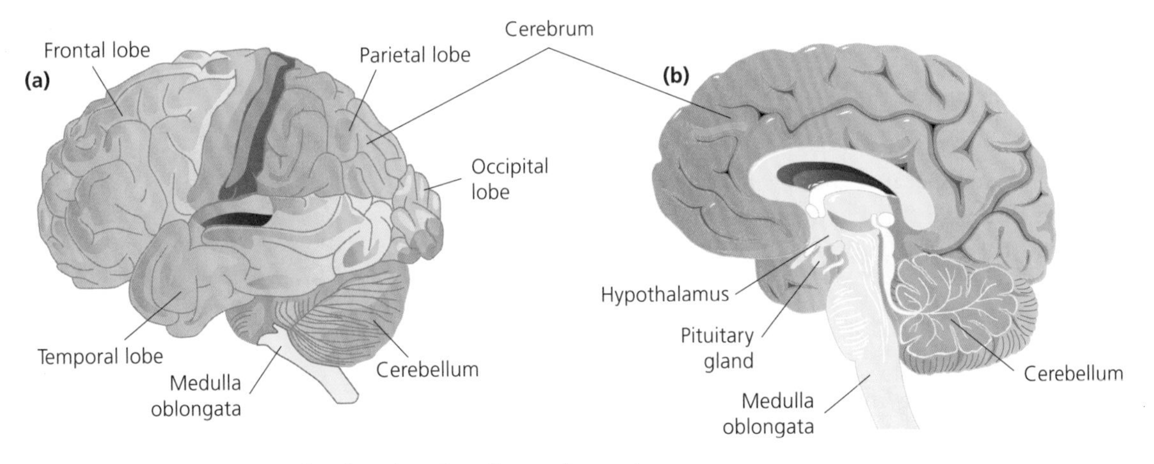

Figure 18.2 The human brain, showing the main parts

Reflex actions

The pathway of a nerve impulse from stimulus to response involves the central nervous system:

stimulus → receptor → sensory neurone → CNS → motor neurone → effector → response

Reflex actions are rapid responses to stimuli that do not involve any conscious thought. The rapid action has survival value because it can protect the body or part of the body from a hazard or reduce damage caused by some external factor such as heat.

Spinal reflexes

REVISED

A knee jerk reflex is an example of a spinal reflex (Figure 18.3). The stimulus is a tap on the knee just below the knee cap. This tap moves a tendon connecting the main thigh muscle (quadriceps) to the lower leg. Stretch receptors in the muscle detect the movement and send an action potential to the spine. This action potential passes straight from the sensory neurone to a motor neurone, which causes the same muscle to contract, pulling the lower leg forwards.

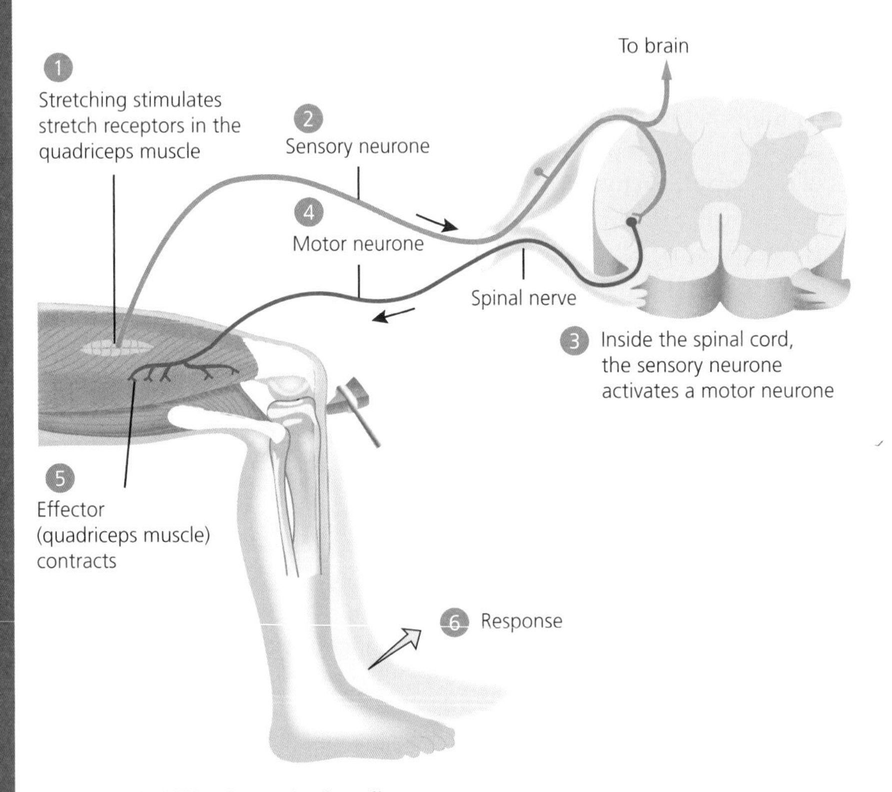

Figure 18.3 **The knee jerk reflex**

This reflex action is unusual because there is no intermediate neurone. In most reflex pathways there is a third neurone connecting the sensory and motor neurones across the central nervous system.

Now test yourself

TESTED

3 Suggest why doctors use the knee jerk reflex to test reflex actions.

Answer on p. 228

Cranial reflexes

REVISED

The blink reflex involves the brain but does not involve any higher thought processes. Stimuli such as a touch or a bright light will cause temporary closure of the eyelid to protect the eye from damage.

Typical mistake

Students often fail to distinguish between the stimulus and the receptor, also between the effector and the response.

Revision activity

Draw another diagram similar to Figure 18.3 to describe the blink reflex.

Coordinated responses

Many responses are actually coordinated by both the nervous system and the endocrine system. One example is the 'flight-or-fight' response seen in mammals. In a dangerous or stressful situation the body is prepared for activity:

- The sensory receptors detect the environmental changes.
- Sensory neurones carry action potentials to the CNS.
- Spinal or cranial reflex actions may bring about very rapid responses.
- Impulses are also conducted to the cerebrum, which uses the association centres to make decisions about how to respond.
- Impulses are sent down the somatic motor neurones to the skeletal muscles to bring about coordinated voluntary movement.
- The hypothalamus is stimulated and sends impulses down the sympathetic part of the autonomic nervous system via the medulla oblongata to bring about a range of changes:
  - The cardiac accelerator nerve carries impulses to the heart, which increase the heart rate and stroke volume.
  - Blood pressure is increased.
  - The breathing rate and depth may rise.
  - Blood vessels to the gut and skin constrict.
  - Blood vessels to the muscles dilate.
  - The adrenal glands are stimulated to release adrenaline, which stimulates the liver to release glucose and maintains the other effects of the sympathetic nervous system. The mechanism of adrenaline action on the cells of its target organs via the release of a second messenger (cAMP) is described in on page 135.

Now test yourself

4 Suggest why the flight-or-fight response is coordinated by both nervous and endocrine systems.

Answer on p. 228

TESTED

Control of heart rate

Heart rate is controlled by both nervous mechanisms and by hormonal mechanisms (Figure 18.4). The heart muscle is myogenic, which means that it can generate its own rhythm. However, this rhythm is overridden by the sinoatrial node (SAN). The SAN initiates waves of excitation that cause the heart to beat.

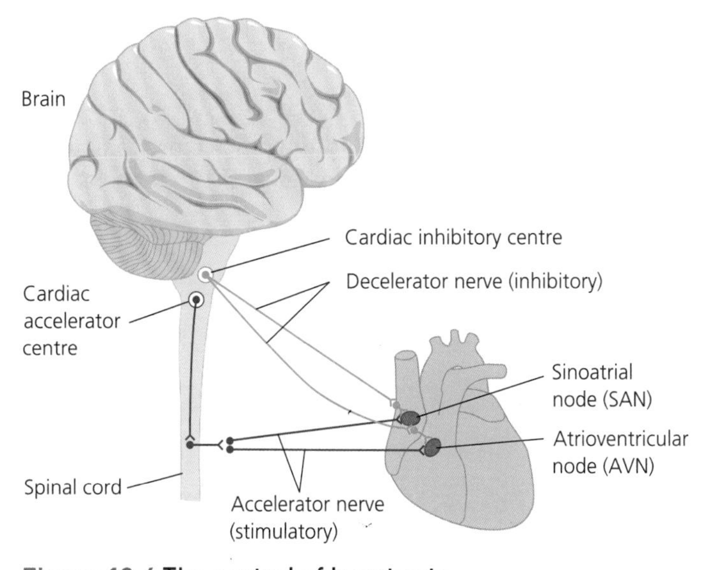

Figure 18.4 The control of heart rate

The rate at which the SAN initiates contractions is affected by two nerves from the cardiovascular centre in the medulla of the brain:

- The decelerator nerve (vagus nerve) releases acetylcholine, which reduces the heart rate.
- The accelerator nerve releases noradrenaline, which is similar to adrenaline and increases the heart rate.

The heart also responds to hormones in the blood, particularly to adrenaline.

Muscle types

There are three types of muscle in the body. Each is different in structure and function (Table 18.3).

Table 18.3

Voluntary muscle	Involuntary muscle	Cardiac muscle
Striated, multinucleate, organised in parallel bundles of myofibrils Fibres all parallel	Smooth muscle, consisting of individual cells	Organised as single cells joined by intercalated discs Parallel myofibrils similar to striated muscle Cross bridges present between fibres
Skeletal muscle attached to bones by tendons	Found in walls of blood vessels, digestive system and airways Contracts slowly	Only found in the heart
Supplied by nerves from the peripheral somatic nervous system	Supplied by nerves from the autonomic nervous system	Supplied by nerves from the autonomic nervous system

Now test yourself

TESTED

5 Suggest why cardiac muscle has cross bridges between the fibres.

Answer on p. 228

Skeletal muscle

Movement requires the coordinated action of voluntary muscles (skeletal muscle). Skeletal muscle can contract (get shorter) but cannot elongate again without the action of an **antagonistic** muscle. Therefore, skeletal muscles work in antagonistic pairs. Movement is achieved by coordinated action of the antagonistic muscles — when one muscle contracts to get shorter the other relaxes and is pulled out or extended.

Antagonistic refers to pairs of muscles that oppose each other.

The muscles are stimulated by motor neurones leading from the motor area of the brain. One motor neurone is attached to one or more muscle fibres — called a motor unit. The neurone is attached to the muscle fibre by a motor end plate or neuromuscular junction. This works just like a synapse to stimulate the muscle fibre membrane:

- Activated by action potentials
- Releases acetylcholine by exocytosis into cleft
- Diffusion of transmitter across cleft
- Receptor sites on the post junction membrane open sodium channels to produce an action potential in the membrane
- Muscle contracts

Exam tip

When asked to compare or contrast in an exam question you can draw a table in the answer space — a table is a quick and easy way to make comparisons.

Now test yourself

TESTED

6 Explain why muscles are always found in antagonistic pairs.

Answer on p. 229

Skeletal muscle is banded or striated, so it is often called striated muscle. The **striations** are produced by the organisation of the actin and myosin protein filaments inside the muscle fibres (Figure 18.5). Each fibre contains many nuclei — it is known as multinucleate. The fibres are surrounded by a membrane known as the sarcolemma, which is equivalent to the plasma membrane of cells. The cytoplasm is specialised and is called sarcoplasm. It contains many mitochondria to supply energy to the contraction process. The endoplasmic reticulum is also specialised to form sarcoplasmic reticulum, which is involved in controlling contractions.

Striations are banding patterns created by the protein filaments in the fibre.

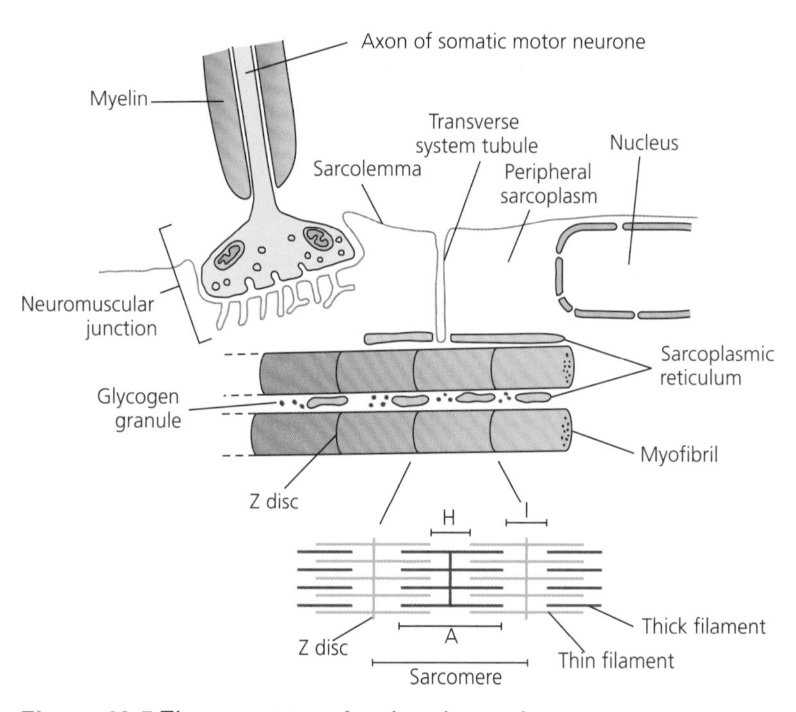

Figure 18.5 The structure of striated muscle

Striations and contraction

REVISED

The striations are caused by the arrangement of protein filaments in the fibre. The dark band (A band) is made up of thicker myosin filaments, which are held together by the M line. The lighter band (I band) is made up of thinner actin filaments, which are held together by the Z line. The distance between two Z lines is called a sarcomere and this is the active unit of a muscle. The actin and myosin filaments overlap.

Contraction is brought about by the actin and myosin filaments sliding past one another. This moves the Z lines closer together so that the sarcomeres get shorter, as does the whole fibre. During contraction the banding pattern changes:

- The I band gets narrower.
- The H zone (the region of the A band where there is no overlap with the actin) gets narrower.

Now test yourself

7 Explain why the I band and H zone get narrower during contraction.

Answer on p. 229

TESTED

The mechanism of contraction

REVISED

The myosin filaments have specialised regions called heads. The **myosin heads** can bridge the gap between the actin and the myosin. When a myosin head attaches to a special binding site on the actin it changes shape pulling the actin past the myosin — making the filaments slide. The activity of myosin is controlled by calcium ions.

The **myosin head** is the part of the myosin filament that can move and attach to the actin filament.

When the muscle is stimulated calcium is released from the sarcoplasmic reticulum. The calcium binds to a protein (troponin), which is associated with the actin filament. The troponin changes shape to move another protein (tropomyosin), exposing the myosin binding sites on the actin.

Energy for contraction

REVISED

The myosin head is an ATPase. When the actin and myosin are locked together ATP joins the myosin head. It is broken down, releasing energy that separates the actin and myosin and moves the myosin head back to its starting position. When the myosin rebinds to the actin and slides the actin forward the ADP and P_i are released (Figure 18.6).

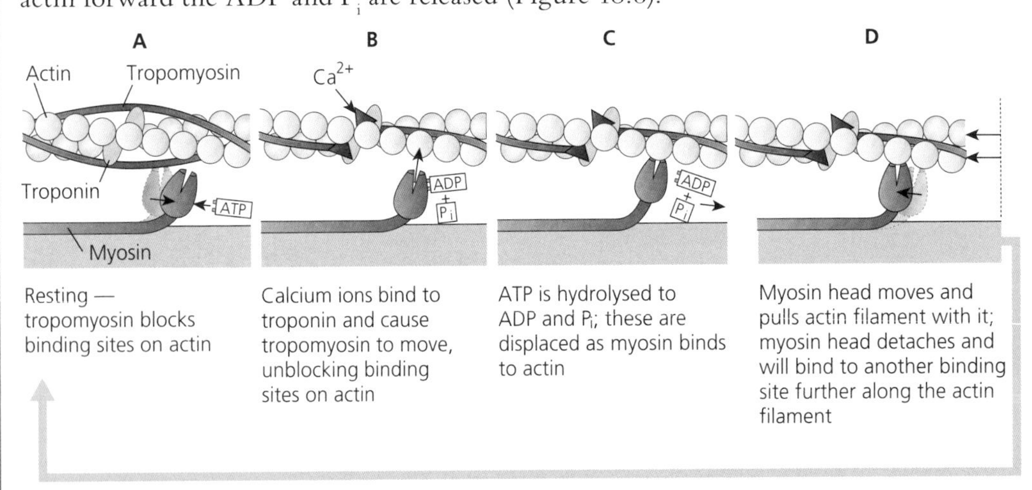
A
B
C
D
Actin
Tropomyosin
Ca^{2+}
Troponin
ATP
ADP
P_i
Myosin
Resting — tropomyosin blocks binding sites on actin
Calcium ions bind to troponin and cause tropomyosin to move, unblocking binding sites on actin
ATP is hydrolysed to ADP and P_i; these are displaced as myosin binds to actin
Myosin head moves and pulls actin filament with it; myosin head detaches and will bind to another binding site further along the actin filament

Figure 18.6 The role of ATP during contraction

Typical mistake

The energy released when ATP is broken down is used for the recovery stroke, not for the power stroke.

ATP supplies are maintained by a reaction involving creatine phosphate (CP). Muscle tissue contains CP, which can transfer phosphate to ADP to make ATP. Muscle tissue also contains glycogen, which can be used anaerobically or aerobically in respiration.

Revision activity

Write a list of all the key terms used in this chapter. Write the meaning of the key term next to each one.

Exam practice

1 (a) Complete the table describing the functions of parts of the human brain. [4]

Area of brain	Function
Cerebrum	
	Control of heart rate
	Thermoregulation
Cerebellum	

(b) Discuss the role of the autonomic nervous system in controlling heart rate. [4]

2 A student investigated the effects of auxin and gibberellin on plant growth. She took ten shoots growing in a suitable medium. Five shoots were coated with gibberellin paste, and five were coated with auxin paste. She measured the length of the side (lateral) shoots each day for 16 days. Her results are shown in the graph.

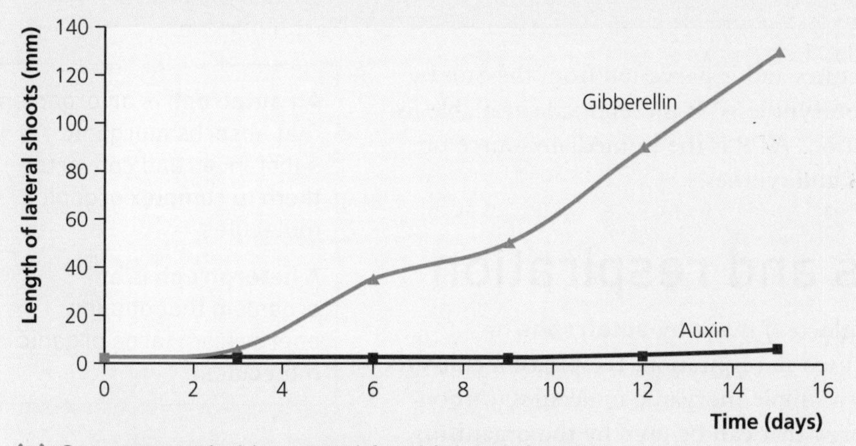

(a) Suggest a suitable control for this experiment. [2]

(b) Suggest why the student used five shoots in each sample. [3]

After 15 days the shoots with auxin had grown to 6 mm and the shoots with gibberellin had grown 130 mm. The student concluded that gibberellin has a greater effect on the growth of the lateral shoots than auxin.

(c) Calculate the % increase in growth at 15 days of shoots with gibberellin applied compared with shoots with auxin applied. [2]

(d) Suggest why the student's conclusion may not be accurate. [2]

3 **(a)** The cerebellum in the brain is concerned with the control and coordination of movement and posture. Suggest why the cerebellum of a chimpanzee is relatively larger than the cerebellum of a cow. [2]

(b) Describe the control and mechanism of muscular contraction. [9]

(c) State *three* differences between cardiac muscle and striated muscle. [3]

Answers and quick quiz 18 online

ONLINE

Summary

By the end of this chapter you should be able to:

- explain why plants need to respond to their environment to avoid herbivory and abiotic stress
- describe practical investigations into tropisms and the effect of plant hormones on growth
- describe the roles of plant hormones
- evaluate the experimental evidence for the role of auxins in the control of apical dominance and gibberellin in the control of stem elongation and seed germination
- describe how plant hormones are used commercially
- outline the organisation of the nervous system in mammals
- outline the organisation and roles of the autonomic nervous system
- describe the gross structure of the human brain and outline the functions of the cerebrum, cerebellum, medulla oblongata, hypothalamus and pituitary gland
- describe the coordination of reflex actions
- describe the coordination of responses by both nervous and endocrine systems, including the flight-or-fight response in mammals and the roles of first and second messengers
- describe the effects of hormones and nervous mechanisms on heart rate
- describe the structure of mammalian muscle and explain the sliding filament model of muscular contraction, including the role of ATP
- outline the structural and functional differences between voluntary, involuntary and cardiac muscle

19 Energy for biological processes: photosynthesis

Energy is essential to living processes and is harvested from the Sun by the biochemical pathways in photosynthesis. It is then made available to the cells through cellular respiration. ATP is the immediate source of energy for biochemical processes and synthesis.

An **autotroph** is an organism that absorbs inorganic substances and converts them to complex organic molecules.

A **heterotroph** is an organism that obtains energy from large organic molecules.

Photosynthesis and respiration

The organic molecules (such as glucose) made by **autotrophs** or consumed by **heterotrophs** are used in respiration. Respiration converts the complex organic molecules to simple inorganic molecules (carbon dioxide and water) releasing energy that can be used by the organism.

Respiration in plants and animals depends upon two products of photosynthesis:

- The complex organic molecules that are broken down in respiration are formed from the products of photosynthesis.
- The oxygen released as a by-product in photosynthesis enables the complex organic molecules to be fully broken down in aerobic respiration.

Typical mistake

Many students forget that plants still need to respire — they photosynthesise during the day but they respire all the time.

Chloroplasts

Photosynthesis takes place in chloroplasts. These are specialised organelles found inside plant cells (Figure 19.1). Photosynthesis has two stages known as the light-dependent stage and the light-independent stage.

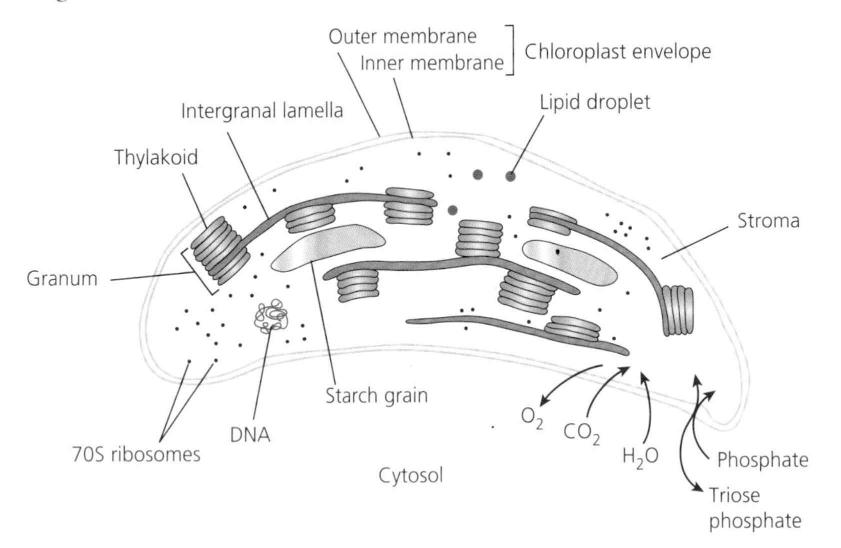

Figure 19.1 A chloroplast and the exchange that occurs with the cytosol

- The chloroplast has an inner and an outer membrane, forming the chloroplast envelope.
- The chloroplast contains many membranes that result from infoldings of the inner membrane and produce a large surface area for the reactions associated with the light-dependent stage. These membranes

are arranged into discs called **thylakoids**. The thylakoids are arranged into stacks called **grana** (singular granum). The grana are connected by membranes called lamellae.

- The thylakoids contain many photosynthetic pigments arranged with enzymes and coenzymes to enable the light-dependent stage to proceed effectively in their membranes.
- The thylakoids are surrounded by a fluid called stroma, which contains all the enzymes needed for the light-independent stage of photosynthesis.
- Chloroplasts also contain their own DNA, allowing them to manufacture the enzymes and proteins specific to photosynthesis.

Thylakoids are discs of specialised membranes within the chloroplast.

A **granum** is a stack of thylakoids.

Revision activity

Sketch a diagram of a chloroplast and annotate the diagram with information about the role of each structure in photosynthesis.

Photosynthetic pigments

Photosynthetic pigments are large organic molecules that absorb light and convert it to a form of energy that can be used in photosynthesis.

There is a range of photosynthetic pigments that each absorb a range of wavelengths. Each pigment has a specific peak of absorption. Chlorophyll a is the main pigment. It absorbs red and blue light but reflects green light. It is known as the **primary pigment**. It is blue-green and has two forms. P_{680} has a peak of absorption at 680 nm; P_{700} has a peak of absorption at 700 nm.

A **photosynthetic pigment** is a pigment that absorbs light energy over a range of wavelengths.

The **primary pigment** is the chlorophyll a found in the reaction centre.

Typical mistake

Many students seem to think that these versions of chlorophyll a only absorb light at the one wavelength — don't forget that the figure refers to the peak of absorption.

Other pigments are called **accessory pigments**. These absorb additional wavelengths that are not absorbed well by chlorophyll a. They include:

- chlorophyll b, which is yellow-green
- carotene, which looks orange and absorbs blue light
- xanthophyll, which looks yellow and absorbs blue light

The **accessory pigments** are pigments that absorb energy and pass it to the primary pigment.

Now test yourself

TESTED

1 Explain why chlorophyll looks green.

Answer on p. 229

Now test yourself

TESTED

2 Suggest why deciduous leaves turn yellow in autumn.

Answer on p. 229

Light absorption

The accessory pigments are arranged in a funnel-shaped, light-harvesting apparatus in the membranes of the thylakoids. They absorb light and pass its energy down the funnel. Chlorophyll a is found at the base of the funnel, in the reaction centre. Electrons are excited by light energy hitting the molecule or by energy passed on from the accessory pigments. The excited electrons are more energetic and move to higher energy orbits. The energy carried by the electrons can then be used in the light-dependent stage of photosynthesis.

Exam tip

Remember that light absorption is all about harvesting the Sun's energy — the light is converted to chemical energy in the form of excited electrons.

Revision activity

Draw a mind map with 'photosynthetic pigment' at the centre. Include a definition and the roles of primary and accessory pigments.

The light-dependent stage

The light-dependent stage of photosynthesis occurs only in the presence of light. It occurs in the membranes of the thylakoids. Each membrane contains a lot of light-harvesting funnels, each with a reaction centre at the base. There are two types of reaction centre, each containing a photosystem with a slightly different version of chlorophyll a:

- photosystem 1 contains chlorophyll a (P_{700})
- photosystem 2 contains chlorophyll a (P_{680})

The remaining enzymes, coenzymes and electron carriers are located in the membrane close to the reaction centres.

Photolysis

REVISED

Photolysis means splitting by light. One of the enzymes is used to split water to produce H^+ ions and electrons:

$$2H_2O \rightarrow 4H^+ + 4e^- + O_2$$

The hydrogen ions and electrons are used during the light-dependent stage. Oxygen is released as a by-product. Photolysis occurs in association with photosystem 2.

Cyclic photophosphorylation

REVISED

Photophosphorylation is the combination of ADP and a phosphate group in the presence of light. In photosystem 1 (P_{700}) the energy from electrons excited by light can be used to generate ATP. Each electron passes back to the same pigment system, so this is called cyclic photophosphorylation (Figure 19.2):

- The excited electron is picked up by an electron carrier.
- It travels down an electron transport chain.
- Energy is released in small amounts and is used to pump hydrogen ions across the thylakoid membrane into the thylakoid disc.
- This builds up a concentration gradient.
- Hydrogen ions diffuse back out of the disc through specialised channels attached to the enzyme ATP synthase.

- The movement of the hydrogen ions provides the energy to combine ADP with P_i, making ATP.
- This process used to manufacture ATP is called chemiosmosis.

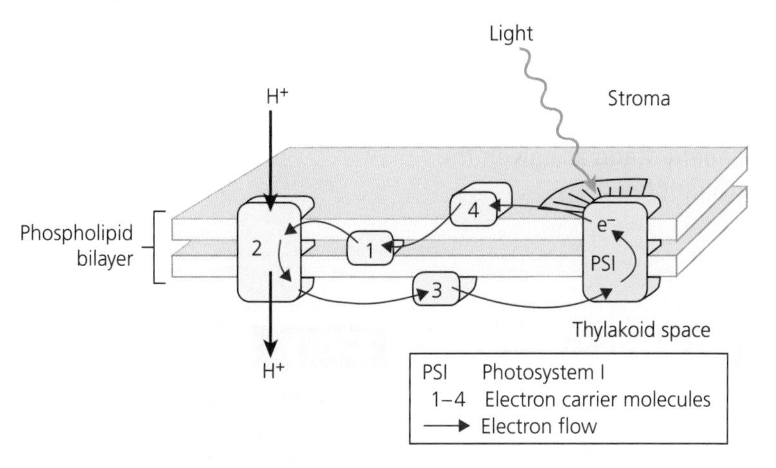

Figure 19.2 Cyclic photophosphorylation

Non-cyclic photophosphorylation

REVISED

The excited electron from photosystem 1 may take another route. When the electron is not returned to photosystem 1 this is called **non-cyclic photophosphorylation** (Figure 19.3).

Non-cyclic photophosphorylation uses the energy from an electron that has been excited by light absorption but does not return it to its original position.

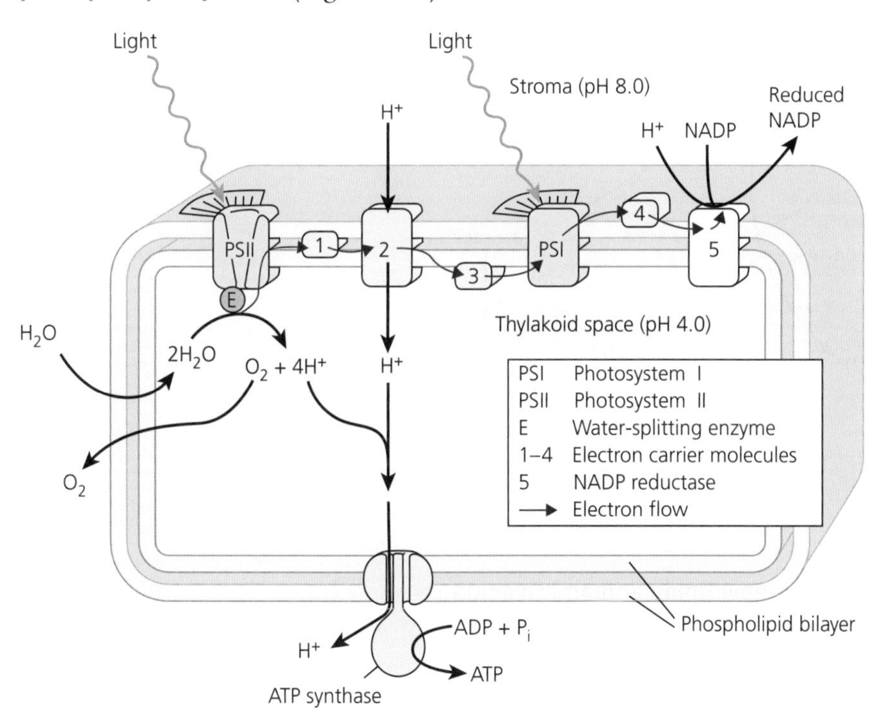

Figure 19.3 Non-cyclic photophosphorylation

- The electron from photosystem 1 is combined with H^+ (from photolysis) and NADP to make reduced NADP. This leaves photosystem 1 unbalanced because it is missing an electron.
- An electron from photosystem 2 (P_{680}) is also excited.
- This electron is picked up by an electron carrier and taken down an electron transport chain to photosystem 1. This replaces the electron lost from photosystem 1.
- The electron transport chain is also associated with the production of ATP through chemiosmosis.

Exam tip

This topic seems very complicated. However, if you view the process as a flow diagram, it is easy to follow. Remember that no additional biochemical detail is needed.

- However, photosystem 2 is left unbalanced because it is missing an electron. This is replaced by the electron released from photolysis.
- The products of the light-dependent stage are ATP and reduced NADP. These are used in the light-independent stage.

Revision activity

Draw a flow diagram of the reactions involved in the light-dependent stage (this diagram is often referred to as the Z-scheme).

Now test yourself

TESTED

3 Explain why the reactions of the light-dependent stage are given the names cyclic and non-cyclic photophosphorylation.

Answer on p. 229

The light-independent stage — the Calvin cycle

Exam tip

The light-independent stage of photosynthesis can occur in the dark — but it uses the products of the light-dependent stage.

The light-independent stage of photosynthesis occurs in the stroma of the chloroplasts. The stroma contains all the enzymes needed for fixing carbon dioxide to produce complex organic molecules (Figure 19.4).

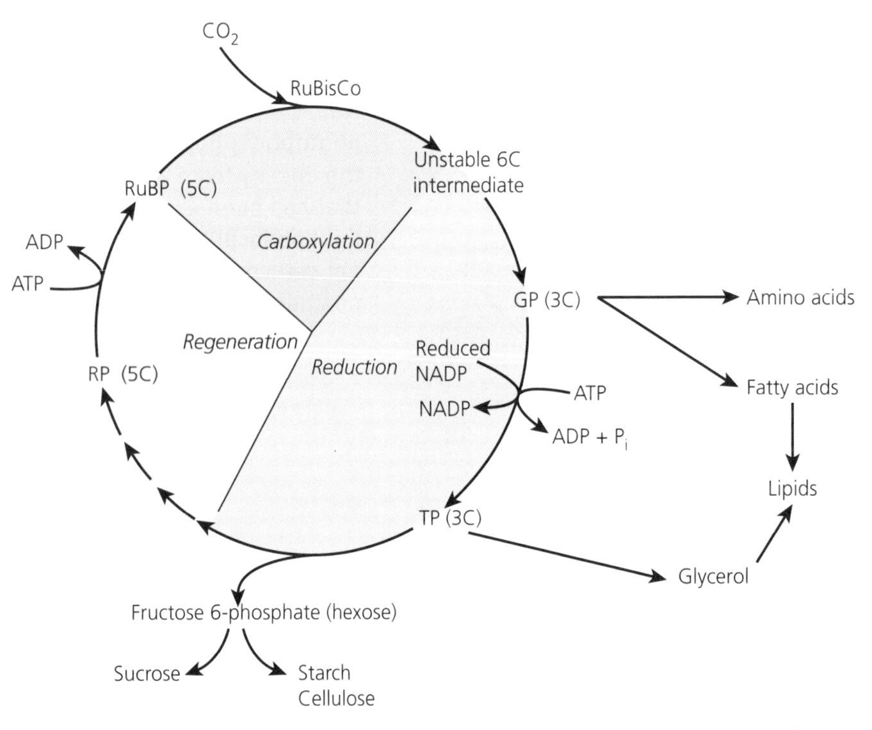

Figure 19.4 The Calvin cycle, also showing how lipids and amino acids can be produced

Carbon dioxide absorbed from the atmosphere is combined with ribulose bisphosphate (RuBP) in a process called fixing. The enzyme ribulose bisphosphate carboxylase (RuBisCO) speeds up this process. An unstable 6-carbon compound is produced, which rapidly breaks down to two molecules of glycerate 3-phosphate (GP). The GP is then converted to triose phosphate (TP) by reduction using the reduced NADP and ATP from the light-dependent stage. The TP can then be used to manufacture the large organic molecules needed by the plant. These include:

- carbohydrates such as glucose
- amino acids
- lipids

However, most of the TP is recycled to produce more RuBP. This process requires more ATP.

Exam tip

Remember that no additional biochemical detail is needed.

Limiting factors

The rate of photosynthesis can be affected by a number of **limiting factors**. Only one factor can limit the rate of photosynthesis at any one time, so these factors interact with one another (Figure 19.5).

A **limiting factor** is something that can limit the rate of a reaction if there is a limited supply.

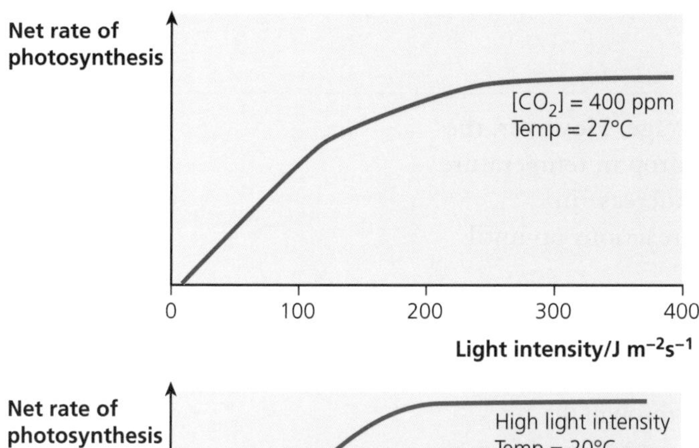

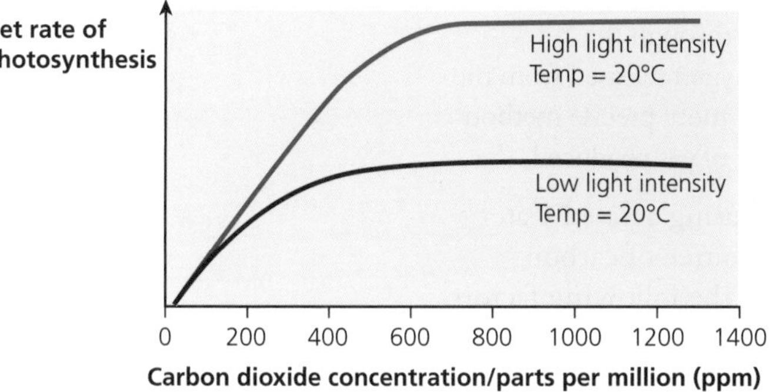

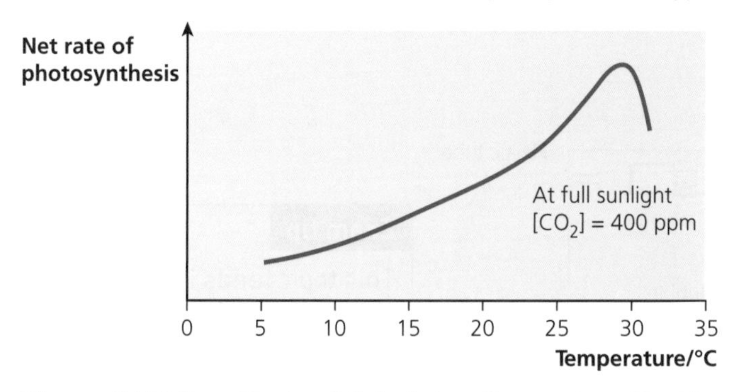

Figure 19.5 The effects of light intensity, carbon dioxide concentration and temperature on the rate of photosynthesis

Carbon dioxide concentration

REVISED

If the CO_2 concentration rises the rate of photosynthesis increases until another limiting factor prevents any further increase. If the concentration of CO_2 drops, less CO_2 can be fixed, so less GP is made. This means that less TP and less glucose is made. The concentration of RuBP rises.

Exam tip

Remember to refer to the concentration of carbon dioxide.

Light intensity

REVISED

If light intensity increases then the rate of photosynthesis increases until another factor prevents any further increase. A reduction in light intensity slows the light-dependent stage. As a result, less ATP and reduced NADP is available to power the Calvin cycle. Less GP is converted to TP and less TP can be converted to RuBP, so the concentration of GP and TP remains high while the concentration of RuPB drops. This means that less carbon dioxide can be fixed.

Typical mistake

Many students lose marks by referring to 'light' changing rather than 'light intensity' changing.

Now test yourself

TESTED

4 Explain why reduced light intensity leads to an increase in the concentration of GP and a reduction in the amount of glucose produced.

Answer on p. 229

Temperature

REVISED

Temperature has little effect on the light-dependent stage. However, the light-independent stage is controlled by enzymes. A drop in temperature reduces the rate of enzyme-controlled reactions. An increase in temperature increases the rate of enzyme-controlled reactions up until the enzymes are denatured.

Measuring the rate of photosynthesis

REVISED

The easiest way to investigate the rate of photosynthesis is through the production of oxygen. It is possible to count bubbles of oxygen released from the cut stem of an aquatic plant such as *Elodea*. Alternatively, a more precise method is to use a photosynthometer, which measures the volume of gas produced.

The apparatus should be set as shown in Figure 19.6 using aerated water that contains extra sodium hydrogen carbonate as a source of carbon dioxide for the plant. After a period of equilibration, the following factors can be investigated:

- Changing temperature of the water bath.
- Changing CO_2 concentration by adding sodium hydrogen carbonate to the water.
- Changing light intensity by moving the lamp.

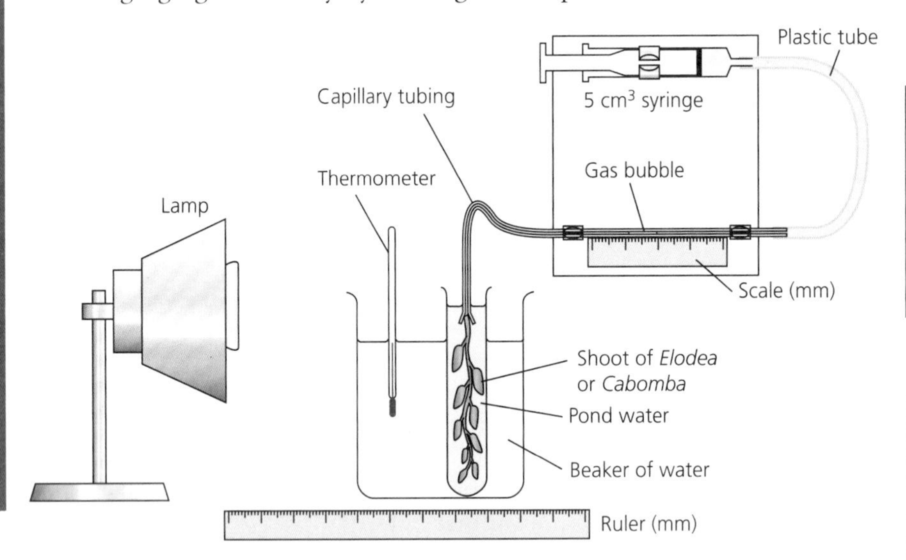

Figure 19.6 A photosynthometer

Exam tip

This topic lends itself to testing How Science Works — interpreting and evaluating the methods and results of investigations.

Revision activity

Write a list of all the key terms used in this chapter. Write the meaning of the key term next to each one.

Now test yourself

TESTED

5 Explain why:
(a) the water used should be aerated
(b) the water should contain extra sodium hydrogen carbonate
(c) the apparatus should be left for 5 minutes before readings are made
(d) using a photosynthometer is more precise than counting bubbles

Answer on p. 229

Exam practice

1 (a) Define the term photosynthetic pigment. [1]
(b) Name the primary photosynthetic pigment and two accessory pigments. [3]
(c) Explain why most leaves contain more than one photosynthetic pigment. [3]
(d) Explain why many aquatic plants living at greater depths are dark in colour or almost black. [3]

2 A student investigated the effect of different coloured light on the rate of photosynthesis. He placed different coloured filters between a light and a piece of pondweed (*Elodea*) and counted the number of bubbles released from the pondweed in 2 minutes. His results are shown in the table.

Colour of filter	Number of bubbles counted in 2 minutes
Black	12
Blue	78
Green	18
Yellow	32
Red	96

(a) (i) State which colour of light enabled the most rapid rate of photosynthesis. [1]
(ii) Explain why there was very little oxygen released with the black filter. [2]
(iii) Explain why there was little oxygen released with the green filter. [2]
(b) The student expected there to be no oxygen released with the black filter in place. Suggest two reasons why some bubbles were observed when the black filter was in place. [2]
(c) Suggest two ways in which the student could have improved this investigation. [2]

3 (a) (i) What is meant by the term *photophosphorylation*. [2]
(ii) Describe how ATP is produced through non-cyclic photophosphorylation. [4]
(b) Describe why oxygen is released as a by-product of the light-dependent stage of photosynthesis. [3]

4 (a) Identify *two* products of the light-dependent stage of photosynthesis that are used in the Calvin cycle. [2]
(b) Describe how carbon dioxide is fixed. [2]
(c) Explain the effects of low carbon dioxide concentration on the rate of photosynthesis. [3]

Answers and quick quiz 19 online

ONLINE

Summary

By the end of this chapter you should be able to:

- describe the interrelationship between respiration and photosynthesis
- explain the structure of a chloroplast
- define the term photosynthetic pigments and explain their importance in photosynthesis
- state that the light-dependent stage takes place in thylakoid membranes and that the light-independent stage takes place in the stroma
- outline how light energy is converted to chemical energy (ATP and reduced NADP) in the light-dependent stage
- outline how carbon dioxide is fixed and how the products of the light-dependent stage are used in the light-independent stage
- state that TP can be used to make carbohydrates, lipids and amino acids but that most TP is recycled to RuBP
- discuss limiting factors in photosynthesis
- describe the effect on the rate of photosynthesis and on levels of GP, RuBP and TP of changing carbon dioxide concentration, light intensity and temperature
- describe how to investigate experimentally the factors that affect the rate of photosynthesis

20 Energy for biological processes: respiration

The need for respiration

Respiration is a series of chemical reactions that take place inside living cells. The reactions release energy from complex organic molecules in a controlled way, so that the energy can be used by the cell.

Many processes inside cells require energy and the immediate source of energy is ATP. The reactions of respiration produce ATP so that it can be used to drive processes such as active transport and energy-requiring metabolic reactions such as building large organic molecules (proteins, DNA etc.).

Respiration is a series of reactions that release energy from substrate molecules inside cells.

Mitochondrial structure

The mitochondria are highly adapted to enable them to carry out their role in respiration. These adaptations include:

- two membranes separated by an inter-membrane space
- inner membrane folded into cristae to create a larger surface area
- ATP synthase molecules seen as stalked particles on the cristae
- presence of electron carrier molecules in the inner membrane
- presence of the enzymes needed for the Krebs cycle in the matrix
- presence of loops of DNA and ribosomes that enable the mitochondrion to manufacture the enzymes and other proteins specific to aerobic respiration

Revision activity

Draw a diagram of a mitochondrion and annotate the diagram with notes about how the features enable the mitochondrion to carry out its role in respiration.

The biochemistry of respiration

Aerobic respiration is a complex process. It takes place in four stages:

- glycolysis
- the link reaction
- the Krebs cycle
- oxidative phosphorylation

These are described in outline.

Aerobic respiration is the release of energy from substrate molecules using oxygen.

Revision activity

Using a large sheet of paper draw the outline of a mitochondrion and superimpose a flow diagram showing how the four stages of respiration link to each other. Leave space to add further detail to your flow diagram later.

Exam tip

Remember that you are not expected to know any additional biochemical detail.

Glycolysis

REVISED

Glycolysis occurs in the cytoplasm of every living cell. It is the breakdown of glucose to smaller 3-carbon molecules called pyruvate and does not require oxygen — it is an anaerobic process.

Glycolysis means splitting sugar.

The chain of reactions includes the following stages:

- **Phosphorylation** of glucose occurs by addition of two phosphate groups from two ATP molecules. This produces hexose bisphosphate.
- The hexose bisphosphate splits into two triose phosphate molecules.
- **Oxidation** of the triose phosphate molecules to pyruvate is achieved by removal of hydrogen atoms. This enables:
    - reduction of NAD (as the NAD accepts the hydrogen atoms)
    - formation of some ATP by substrate-level phosphorylation. This is where an enzyme is used to combine ADP and a phosphate group using energy released from the substrate molecule.

Phosphorylation means addition of a phosphate group.

Oxidation means increasing the oxidation number, often by removing hydrogen or electrons.

The products of glycolysis are:

- reduced NAD
- ATP
- pyruvate

Typical mistake

Many students forget that glycolysis takes place in the cytoplasm.

Revision activity

Add this detail to your outline flow diagram.

The link reaction

REVISED

When sufficient oxygen is available for aerobic respiration to take place the pyruvate is actively transported into the mitochondrion. The link reaction takes place in the matrix of the mitochondrion (mitochondrial matrix). It includes a number of steps:

- **Decarboxylation** of the pyruvate to produce a 2-carbon acetyl group and carbon dioxide. The carbon dioxide is released for excretion.
- **Reduction** of NAD by hydrogen atoms released from the pyruvate.
- Combination of the acetyl group with coenzyme A to produce acetylcoenzyme A, which carries the acetyl group into the next stage.

Decarboxylation means removal of a group of atoms including carbon.

Reduction means decreasing the oxidation number, often by addition of a hydrogen or electron.

Revision activity

Add this detail to your outline flow diagram.

The Krebs cycle

REVISED

The Krebs cycle takes place in the mitochondrial matrix. It consists of a complex series of steps:

- The acetyl group is released from the acetylcoenzyme A and combined with a 4-carbon compound called oxaloacetate. This forms the 6-carbon compound called citrate.
- The coenzyme A is released to be used again.
- The citrate enters a cycle of reactions that will release its energy in stages.
- The citrate is decarboxylated and dehydrogenated to reform oxaloacetate.
- Decarboxylation produces carbon dioxide, which is excreted.
- **Dehydrogenation** releases hydrogen atoms, which are used to reduce the coenzymes NAD and FAD.
- These reduced coenzymes act as hydrogen carriers to pass the hydrogen onto the next stage.

Dehydrogenation means removal of a hydrogen atom.

During the Krebs cycle some ATP is produced by substrate-level phosphorylation. This is where an enzyme is used to combine ADP and a phosphate group using energy released from the substrate molecule.

Aerobic respiration is outlined in Figure 20.1.

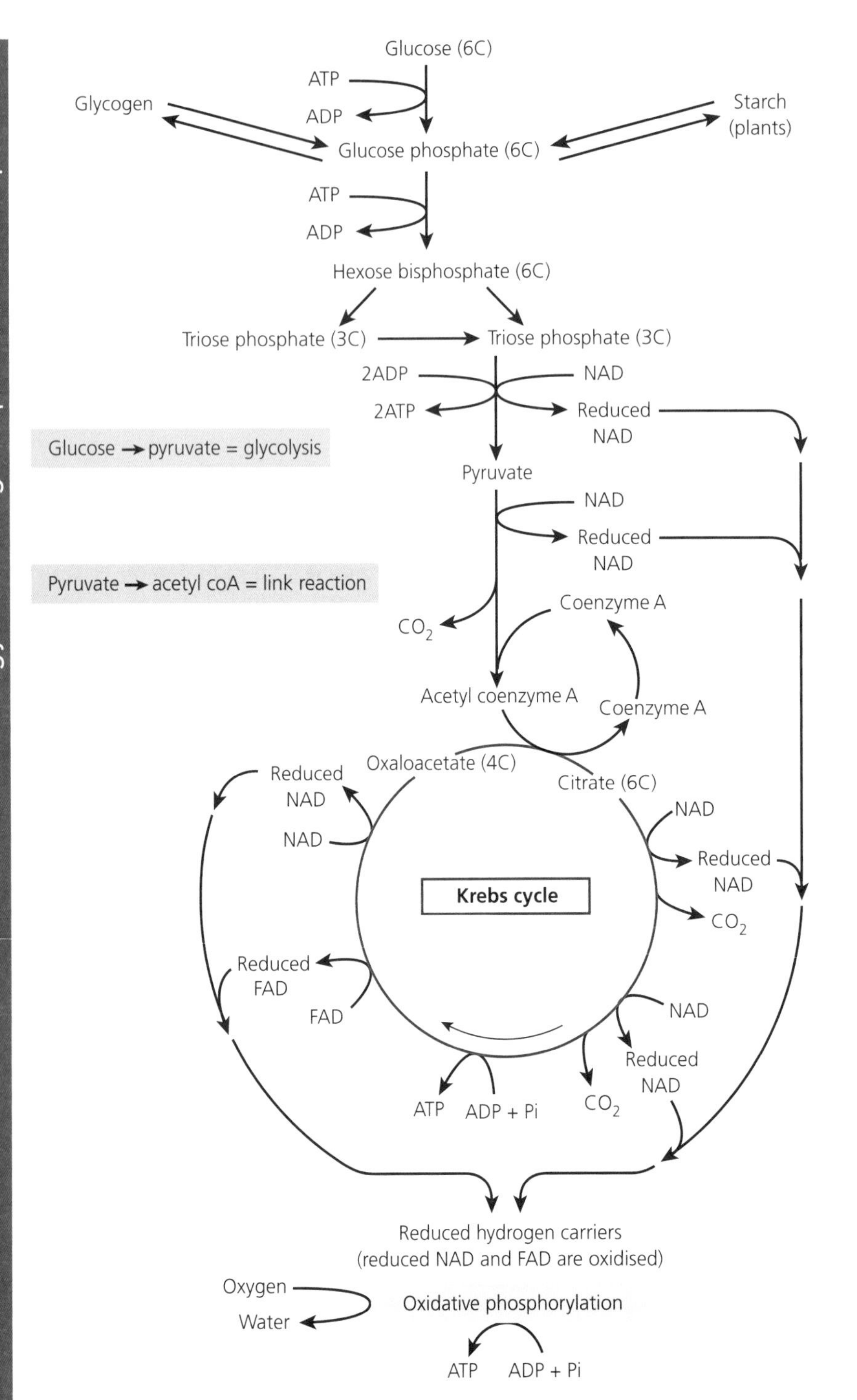

Figure 20.1 A summary of aerobic respiration

Typical mistake

Many students try to learn too many extra names. It can be useful to recognise additional names of intermediates, but it can also become very confusing.

Revision activity

Add this detail to your outline flow diagram.

Exam tip

This topic seems very complicated — and can be made to sound more complicated by using lots of extra names. However, the only names you need to know in the Krebs cycle are oxaloacetate and citrate. If you view the process as a flow diagram, it is easy to follow.

Coenzymes

Coenzymes are complex organic molecules that contribute to enzyme-controlled reactions inside cells. The main role of a coenzyme is to transport products from one reaction into another reaction.

A **coenzyme** is a complex organic molecule used to transfer the products of one reaction to become the substrate in another reaction.

Respiration is a complex series of reactions. Coenzymes are used in a number of places to link the reactions. The coenzymes used include:

- NAD, which is reduced by hydrogen atoms released from the respiratory substrate (usually glucose). The hydrogen atoms can be used in different ways:
  - When oxygen is available they are transported to the inner membrane of the mitochondrion (crista) where they drive an electron transport chain involved in ATP production.

- ○ When oxygen is not available they may be used in reduction reactions, such as reducing pyruvate to lactate in anaerobic respiration.
- Coenzyme A, which binds to a 2-carbon acetyl group in the link reaction and delivers the acetyl group into the Krebs cycle.

Typical mistake

Do not confuse the NAD used in respiration with the NADP used in photosynthesis.

Now test yourself

TESTED

1 Explain why it is important that the coenzyme A is oxidised and released.

Answer on p. 229

Revision activity

Draw a mind map with coenzymes at the centre. Include all the coenzymes from chapters 19 and 20.

Oxidative phosphorylation

REVISED

- The inner membrane of the mitochondrion is folded into cristae.
- The reduced coenzymes carry the hydrogen atoms to the cristae.
- Embedded in the phospholipid bilayer of the cristae is a series of protein complexes called electron carriers. The series of carriers is called an electron transport chain.
- The first carrier is an enzyme that oxidises the reduced coenzyme by removing the hydrogen and splitting it to a proton (H^+) and an electron (e^-). The coenzyme is released, to return into the mitochondrial matrix and accept more hydrogen atoms.
- The proton enters the mitochondrial matrix.
- The electron passes from electron carrier to electron carrier down the carrier chain. As an electron passes between carriers, energy is released.
- At the end of the electron transport chain the electrons are combined with oxygen and hydrogen ions to make water (Figure 20.2). This is why oxygen is so important for respiration to continue.

Oxidative phosphorylation involves the addition of a phosphate group in a process that uses oxidation/reduction reactions.

Typical mistake

Some students try to learn too many names. You do not need to know all the names of the carriers.

Revision activity

Add this detail to your outline flow diagram.

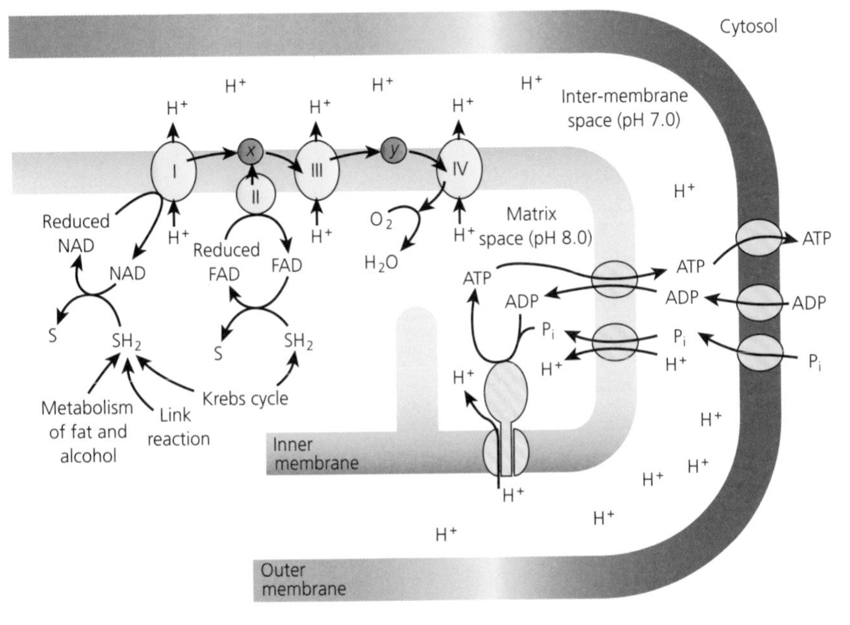

SH_2 = intermediate substances in the link reaction, Krebs cycle and metabolism of fat and alcohol. These substances are dehydrogenated to form reduced hydrogen carriers (reduced FAD and reduced NAD), which are oxidised by complexes I and II.

I, II, III and IV are protein complexes.
x and *y* are electron carrier molecules.

Complexes I, III and IV pump protons; complex II passes electrons from reduced FAD to complex III via electron carrier *x*. Complex II is not a proton pump.

Figure 20.2 A summary of oxidative phosphorylation

Chemiosmosis

- The energy released from the electron as it travels down the electron transport chain is used to pump protons from the mitochondrial matrix into the gap between the two mitochondrial membranes.
- Protons accumulate in the gap to produce a concentration gradient between the gap and the mitochondrial matrix. This is called a proton gradient (it produces a pH gradient and a potential difference across the membrane).
- The proton gradient and the potential difference create a driving force called the proton-motive force, which pushes the protons out into the mitochondrial matrix.
- The protons can only pass through special channel proteins attached to the enzyme ATP synthase.
- As protons flow through the ATP synthase their movement energy is used to produce ATP from ADP and inorganic phosphate.
- The protons return to the mitochondrial matrix and are combined with electrons from the electron transport chain and oxygen to form water. Therefore oxygen is the final electron acceptor in aerobic respiration.

Chemiosmosis is the process involving creation of a proton gradient through the action of an electron transport chain and the subsequent use of the proton motive force to produce ATP. It is used in oxidative phosphorylation and in photophosphorylation.

Revision activity

Add this detail to your outline flow diagram.

Now test yourself TESTED

2 Explain why oxidative phosphorylation and chemiosmosis cannot occur without oxygen.

Answer on p. 229

Anaerobic respiration

Anaerobic respiration takes place in the cytoplasm. If oxygen is not available pyruvate does not enter the mitochondrion and the link reaction, Krebs cycle and oxidative phosphorylation do not occur. Therefore the yield of ATP is much lower (Figure 20.3).

Anaerobic respiration is the release of energy from substrate molecules without using oxygen.

(a) ***mammals***

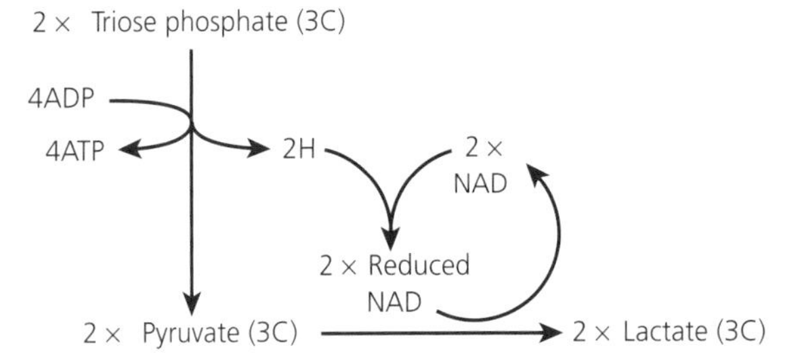

(b) ***yeast***

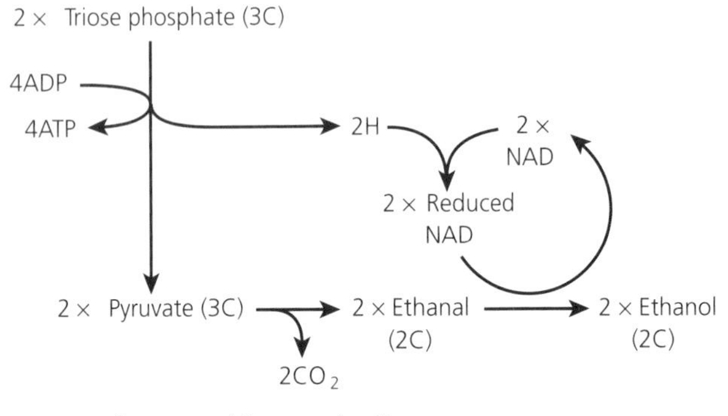

Figure 20.3 A summary of anaerobic respiration

The yield of ATP from anaerobic respiration is just two molecules from each glucose molecule. However, the ability to respire anaerobically allows organisms to survive where there is insufficient oxygen in the environment. It also enables additional muscular contraction even when the blood system cannot supply any more oxygen to the muscles.

In mammals

REVISED

In mammals the pyruvate made in glycolysis is reduced to lactate using the hydrogen from reduced NAD also made in glycolysis. The enzyme lactate dehydrogenase is used.

In yeast

REVISED

In yeast the pyruvate is first decarboxylated to ethanal, releasing carbon dioxide. This involves the enzyme pyruvate decarboxylase. The ethanal is then reduced to ethanol by hydrogen from the reduced NAD made in glycolysis. This step involves the enzyme ethanol dehydrogenase.

Now test yourself

3 Explain why anaerobic respiration releases less ATP from glucose than aerobic respiration.

Answer on p. 229

Respiratory substrates

A **respiratory substrate** is a molecule that can be broken down in respiration to release energy. The following molecules can be used:

- Carbohydrates (glucose, starch and glycogen). Starch and glycogen are first converted to glucose before entering the respiratory pathways.
- Fats. The fatty acids from a fat are broken down to 2-carbon fragments (acetyl group), which enter the Krebs cycle via coenzyme A. Many acetyl groups enter the Krebs cycle from each fat molecule. Therefore a fat molecule can fuel many turns of the Krebs cycle, which means that many molecules of NAD can be reduced by addition of hydrogen atoms. Each reduced NAD or FAD molecule can be used to produce ATP. Fats are therefore energy rich.
- Proteins. These are converted to amino acids, which are then deaminated. The remaining residues are organic acids similar to the intermediates in the Krebs cycle. These enter the Krebs cycle at an appropriate place. Therefore, each amino acid can produce different yields of ATP. The overall energy yield from proteins is similar to that from carbohydrates.

The different energy values of carbohydrates, fats and proteins are shown in Table 20.1.

Table 20.1

Respiratory substrate	Mean energy value/kJ g^{-1}
Carbohydrate	15.8
Protein	17.0
Fat	39.4

A **respiratory substrate** is a molecule that can be broken down in respiration to release energy.

Now test yourself

4 Explain why fats have more energy per gram than carbohydrates

Answer on p. 229

TESTED

Respiratory quotient (RQ)

REVISED

The respiratory quotient is the ratio of carbon dioxide produced to oxygen used during respiration:

$$RQ = \frac{CO_2 \text{ produced}}{\text{oxygen used}}$$

The value of RQ depends on the respiratory substrate (Table 20.2). Under normal conditions there is a mixture of substrates used and the RQ value can be used to estimate the proportion of each substrate used.

Table 20.2

Substrate	Value of RQ
Carbohydrate	1.0
Protein	0.9
Fat	0.7

Measuring the rate of respiration

The rate of respiration can be measured by the uptake of oxygen and release of carbon dioxide. A **respirometer** measures the change in volume of the air inside a chamber containing living tissue such as germinating seeds (Figure 20.4). As oxygen is used the carbon dioxide released is absorbed by the sodium hydroxide. The change in volume represents the volume of oxygen used.

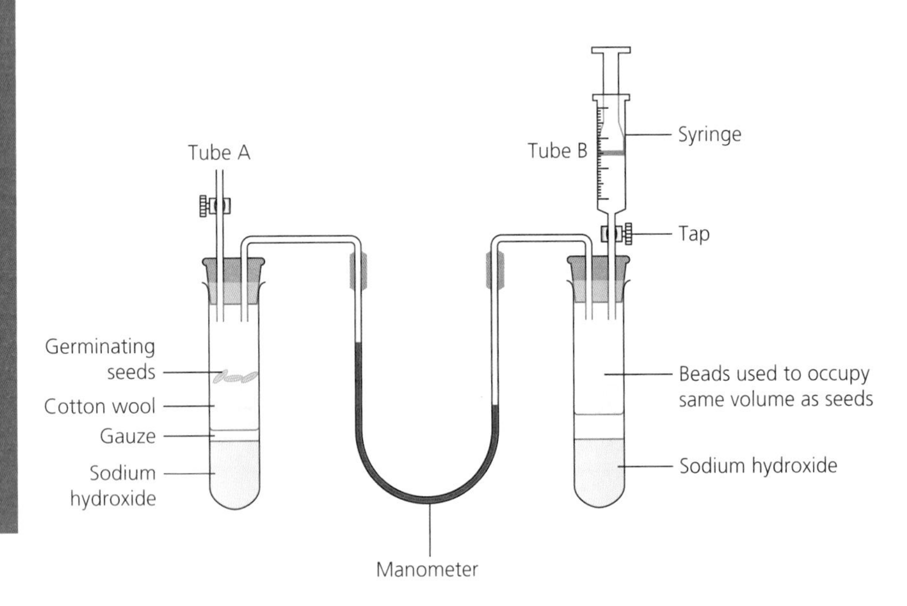

Figure 20.4 A respirometer

- The effect of changing temperature can be investigated by placing the boiling tubes in a water bath at varying temperatures.
- The effect of changing the substrate concentration can be investigated by modifying the apparatus and using yeast with varying concentrations of sugar.
- The effect of changing the substrate can be investigated by supplying the yeast with different substrates.

Revision activity

Write a list of all the key terms used in this chapter. Write the meaning of the key term next to each one.

Exam practice

1 (a) State precisely where in the cell the following stages of respiration take place: glycolysis, the Krebs cycle, oxidative phosphorylation. [3]

(b) Identify the type of reaction occurring at the following stages of respiration:

glucose → glucose 6-phosphate

NAD → NADH

citrate → 5-carbon intermediate

pyruvate → lactate [4]

(c) Describe how ATP is produced during oxidative phosphorylation. [4]

2 (a) Describe and explain two features of a mitochondrion that enable it to perform its role in respiration. [4]

(b) (i) Explain why anaerobic respiration releases only two molecules of ATP compared with 32 molecules released by aerobic respiration. [3]

(ii) Cardiac muscle tends to respire fatty acids rather than glucose. Suggest why the heart muscle is more affected by limited oxygen supply than other types of muscle. [2]

3 (a) Explain what is meant by chemiosmosis. [3]

(b) (i) Some protons leak through the outer mitochondrial membrane. Explain what effect such leakage would have on production of ATP. [3]

(ii) Some of the reduced NAD produced in glycolysis is used to reduce compounds in the cytoplasm. Explain what effect this would have on production of ATP. [3]

4 During oxidative phosphorylation electrons pass down an electron transport chain in the inner mitochondrial membrane.

(a) (i) State the original source of these electrons. [1]

(ii) State the immediate source of these electrons. [1]

(iii) Name the final electron acceptor. [1]

(b) During electron transfer electrons pass from one carrier to another.

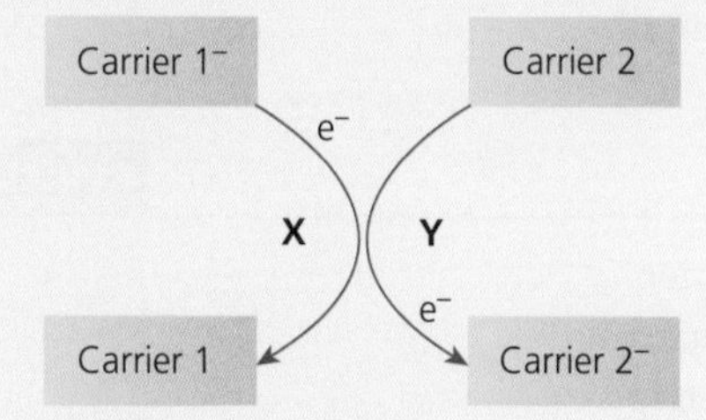

State which letter, X or Y, represents oxidation and which represents reduction. [2]

Answers and quick quiz 20 online

ONLINE

Summary

By the end of this chapter you should be able to:

- outline why all living things need to respire
- explain the structure of the mitochondrion
- state that glycolysis takes place in the cytoplasm and outline the process of glycolysis
- state that in aerobic respiration pyruvate is actively transported into the mitochondria and that the link reaction and Krebs cycle take place in the mitochondrial matrix
- outline the link reaction and Krebs cycle
- explain the importance of coenzymes in respiration
- outline oxidative phosphorylation in the cristae and chemiosmosis
- outline the process of anaerobic respiration in mammals and in yeasts
- explain why anaerobic respiration produces a much lower yield of ATP than aerobic respiration
- investigate aerobic and anaerobic respiration rates in yeast
- define the term respiratory substrate, explain the different relative energy values of carbohydrates, proteins and fats and use the respiratory quotient
- investigate the effect of temperature, substrate concentration and different substrates on the rate of respiration

21 Cellular control

- The control of metabolic reactions determines growth, development and function.
- The genes have an important role in regulating and controlling cell function and development.
- Spontaneous mutations occur to the genes and this can have a major effect on development.

Types of gene mutation

A **gene mutation** is a change to the code in a gene. It involves a change in the sequence of nucleotides in the DNA molecule, resulting in a different base sequence. There is a variety of types of mutation, which can each have a different effect.

A **gene mutation** is a change to the sequence of nucleotides in DNA.

Substitution

REVISED

Substitution is where one base pair is changed to a different base pair. This alters the code for that triplet. It is likely that the new triplet will code for a different amino acid, in which case one amino acid in the polypeptide will be different from the normal sequence. However, due to the degenerate nature of the genetic code, some amino acids have more than one triplet as a code. Therefore the polypeptide may not be affected at all.

Deletion

REVISED

Deletion is where one or more base pairs are deleted from the sequence. This causes frameshift because it can alter every single triplet downstream of the change. If one or two base pairs are deleted then every single triplet downstream of the change is affected. This changes the sequence of amino acids in the polypeptide and the polypeptide or protein produced is very different from its original form. However, if three base pairs (or a multiple of three) are deleted then all the subsequent triplets are the same and the amino acid sequence is not changed as much — the polypeptide simply misses one or more amino acids.

Insertion

REVISED

Insertion is where one or more base pairs are added to the sequence. This has a similar effect to deletion, except that amino acids can be added to the polypeptide, which therefore gets longer.

The effects of gene mutation

The effect that a mutation has depends on the type of mutation. Mutations can be neutral or beneficial, but the majority are harmful. Since most mutated genes are recessive, it is likely that the effects of a mutation will be masked by the production of the normal protein using the dominant allele.

Typical mistake

Many students assume that all mutations are harmful or even lethal.

Neutral mutations

REVISED

Neutral mutations are those that have no effect. This could be because:

- The mutant triplet codes for the same amino acid.
- The mutant triplet changes the amino acid but this has no effect on the function of the polypeptide.
- The mutation occurs in non-coding parts of the DNA.

Beneficial mutations

REVISED

Occasionally a mutation alters the polypeptide such that it works more effectively, but this is unlikely. However, a different version of the gene (an allele) that produces a different version of the polypeptide may become advantageous if the environment changes. This is the basis of variation, which allows natural selection and evolution to occur.

Harmful mutations

REVISED

The majority of mutations are harmful. They alter the structure of the polypeptide such that it works less well or does not work at all.

Now test yourself

TESTED

1 Explain why most mutations are harmful.

Answer on p. 229

Revision activity

Draw a mind map about mutations. Start with the structure of DNA at the centre.

Exam tip

Remember that mutations are rarely lost. They are often recessive and are therefore hidden by the dominant allele. This is important for the success of evolution.

Regulating gene expression

Gene expression is regulated at three points:

- at transcription — known as transcriptional-level regulation
- after transcription — known as post-transcriptional-level regulation
- after translation — known as post-translational-level regulation

Transcriptional-level regulation

REVISED

The lac operon

The lac operon is an example of transcriptional-level regulation. The lac operon is a functional unit of genes in *E. coli*. It contains two genes that code for the structure of proteins and a control mechanism to enable the genes to be switched on and off.

In order to make β-galactoside, RNA polymerase must bind to the promoter region. However, this is usually blocked by a repressor protein. When lactose is present, it binds to the repressor protein, removing it from the operator region so that RNA polymerase can bind. The details are shown in Figure 21.1.

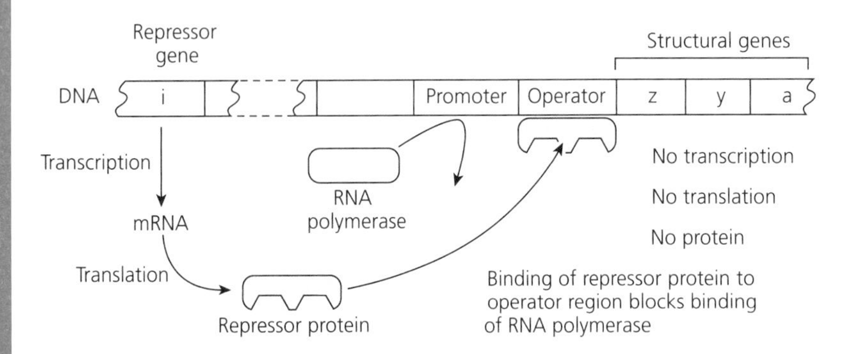

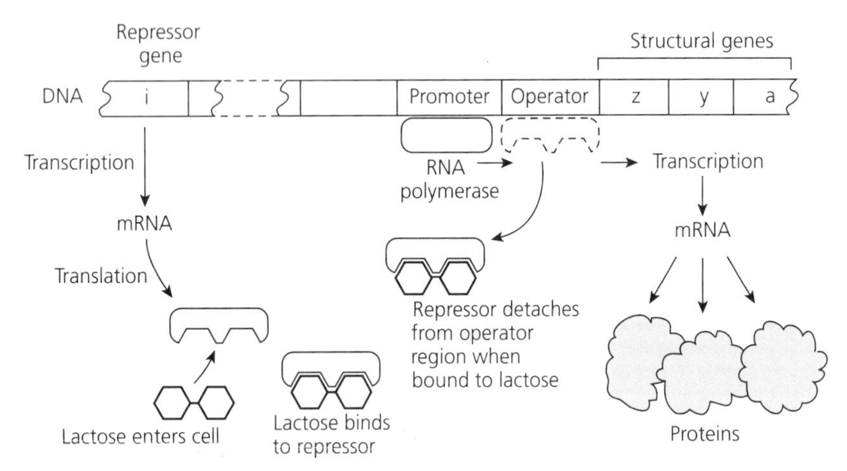

Figure 21.1 **The action of the lac operon (a) when lactose is absent and (b) when lactose is present**

Typical mistake

Some students fail to learn this detail and simply suggest that lactose activates the structural genes.

Now test yourself

2 Suggest why it is useful to the bacterium to be able to switch a gene on and off.

Answer on p. 230

TESTED

Transcription factors in eukaryotes

Transcription factors are proteins that control which genes are turned on or off. Transcription factors bind to specific promoter regions on the DNA in the nucleus. They either help or inhibit the attachment of RNA polymerase to the DNA, so they can either activate or suppress transcription of the gene.

Post-transcriptional-level regulation

REVISED

Post-transcriptional-level regulation occurs when the mRNA produced by transcription is edited. Many genes include additional lengths of DNA known as **introns**. These introns contain sequences that do not contribute to the final protein product. All the DNA is transcribed to produce **primary mRNA**. During post-transcriptional-level regulation these extra sequences are removed from the mRNA to produce **mature mRNA**. This editing and splicing process involves endonuclease enzymes.

The sequences found in the introns might be non-coding or they can code for alternative sequences of amino acids. Through the process of alternative splicing it is possible to produce different versions of mature mRNA from one gene. In effect, this means that it is possible to produce different polypeptides from one gene — the end product depending on how the primary mRNA has been edited.

Introns are lengths of DNA in a gene that do not form part of the code for a protein.

Primary mRNA is the mRNA produced from direct transcription of all the DNA in a gene.

Mature mRNA is the mRNA molecule after the introns have been removed.

Post-translational-level regulation

REVISED

After translation many proteins remain in an inactive form until they are activated. They can be activated by phosphorylation. The response of many target cells to a hormone involves the release of cAMP, which activates the enzyme protein kinase. Protein kinase then converts more ATP to cAMP, releasing two phosphate groups. These phosphate groups are used to phosphorylate and activate other enzymes in the cell. This can form an enzyme cascade because the cAMP produced by protein kinase will activate more protein kinase molecules.

Development of body plans

Homeobox gene sequences

REVISED

Homeobox genes control the basic structure and orientation (body plan) of an organism. This is achieved by controlling the differentiation of cells and parts of the body through switching genes on and off at appropriate times during development.

Homeobox genes are genes that control the body plan of an organism.

The important part of a homeobox gene is a sequence of 180 base pairs known as the homeobox sequence. This codes for a sequence of 60 amino acids, which is found as part of the polypeptide produced from the gene. This sequence of amino acids folds in a way that allows the polypeptide to bind to DNA. The polypeptides produced are transcription factors, which bind to DNA and initiate the transcription of genes.

In animals some of the homeobox genes are arranged into groups called Hox clusters. The homeobox genes in a Hox cluster are called Hox genes. The genes in a Hox cluster are activated in a particular order, which matches the order in which they are expressed along the body from head to tail. As the Hox genes are activated they activate structural genes in a carefully coordinated sequence to ensure that features develop in the correct way.

Typical mistake

Some students do not distinguish between the homeobox gene and the homeobox sequence.

The homeobox genes are very similar in all plants, animals and fungi. This is because they have the same role in each case — they code for transcription factors that need to bind to DNA.

Mitosis and apoptosis

Mitosis is controlled by homeobox and Hox genes. All cells resulting from mitosis contain a full set of genes. Differentiation is achieved by switching the appropriate genes on or off during development.

Apoptosis is a series of carefully controlled biochemical events that leads to orderly cell death. The sequence of events is as follows:

- Enzymes break down the cytoskeleton.
- The cell shrinks and organelles are packed together along with fragments of the chromatin.
- The cell surface membrane breaks up to form cell fragments (vesicles) containing the cell contents.
- The vesicles are taken up by phagocytes and digested.

Apoptosis is programmed cell death.

Apoptosis is an important part of development because some cells need to be killed or removed to ensure correct development. For example:

- T lymphocytes that recognise our own body antigens would attack our own cells if they were allowed to survive and become active.
- During development of the hands and feet tissue grows between the fingers and toes. This must be removed at a later stage.

Apoptosis is controlled by a variety of cell signals. The genes involved in regulating the cell cycle and apoptosis can respond to internal and external cell stimuli such as stress. Cells that detect such stimuli release signalling molecules, which include hormones, cytokines from the immune system, growth factors and nitric oxide.

Revision activity

Write a list of all the key terms used in this chapter. Write the meaning of the key term next to each one.

Exam practice

1 (a) Describe *three* ways in which the structure of RNA and DNA are different. [3]
(b) (i) Explain why the genetic code is said to be a triplet code. [2]
(ii) Explain how some mutations might have no effect on the organism. [4]
(c) Describe how polypeptides are synthesised and explain how the structure of a polypeptide is determined by the sequence of nucleotide bases in the gene. In your answer you should make clear the sequence of events in protein synthesis. [10]

2 (a) The original base sequence in a gene and two mutated versions of the base sequence are shown below:

original sequence: AGTTTCGCCCGT

mutated sequence 1: AGTCTCGCCCGT

mutated sequence 2: AGTTCGCCCGT

(i) Name the type of mutation that has occurred in mutated sequence 1. [1]
(ii) Suggest what effect this mutation may have on the functionality of the polypeptide produced. [2]
(iii) Name the type of mutation that has occurred in mutated sequence 2. [1]
(iv) Suggest what effect this mutation may have on the functionality of the polypeptide produced. [2]
(b) A mutation such as that seen in mutated sequence 2 could occur in the homeobox sequence of a homeobox gene. Explain why this would have far-reaching effects on the organism. [6]

3 (a) State what is meant by apoptosis. [1]
(b) Describe the process of apoptosis. [4]
(c) Explain how apoptosis can contribute to the development of a body plan. [3]

Answers and quick quiz 21 online

ONLINE

Summary

By the end of this chapter you should be able to:

- state that mutations cause changes to the nucleotide sequence and can be beneficial, neutral or harmful
- describe the regulatory mechanisms that control gene expression at transcriptional level, post-transcriptional level and post-translational level
- explain that the genes controlling development of body plans are similar in plants, animals and fungi
- outline how apoptosis can change body plans

Causes of phenotypic variation

Phenotypic variation can be caused by environmental factors and genetic factors.

Environmental factors

REVISED

Environmental factors are those caused by the environment and not through the genes. These include:

- diet in animals — this will affect body size and mass
- exposure to light in plants — limited exposure to light causes chlorosis, where the leaves turn yellow or white and the plant cannot photosynthesise, because the plant is unable to make chlorophyll
- language and dialect spoken
- scarring or loss of limbs

Genetic factors

REVISED

During replication of the DNA to form chromatids the copying of the nucleotide sequence may be inaccurate. This is a mutation. If it occurs in the cells about to undergo meiosis it produces variation.

Meiosis

Genetic variation can also be created by meiosis. When chromatids cross over they exchange lengths of DNA (Figure 22.1). Where this occurs between non-sister chromatids it produces new combinations of **alleles**.

The way the bivalents orientate on the equator during metaphase 1 is random. This means that either the maternal or the paternal chromosome of a bivalent may face either pole. Therefore the combination of maternal and paternal chromosomes migrating to either pole is random. This is called independent assortment of homologous chromosomes (Figure 22.2).

An **allele** is an alternative form of a gene. Alleles are found at the same position on homologous chromosomes. In most cases an organism has two copies of every gene. These could be identical copies or alleles.

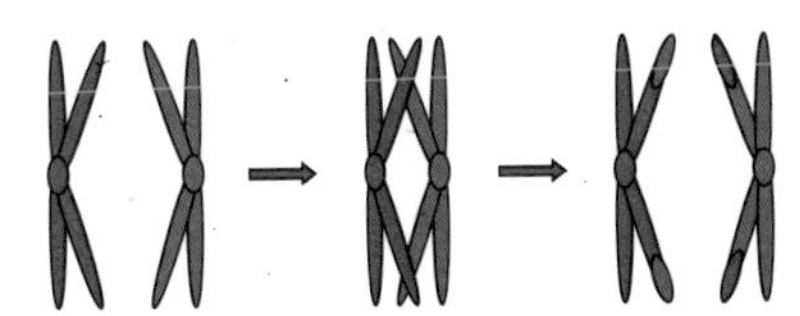

Figure 22.1 The exchange of genetic material by crossing over

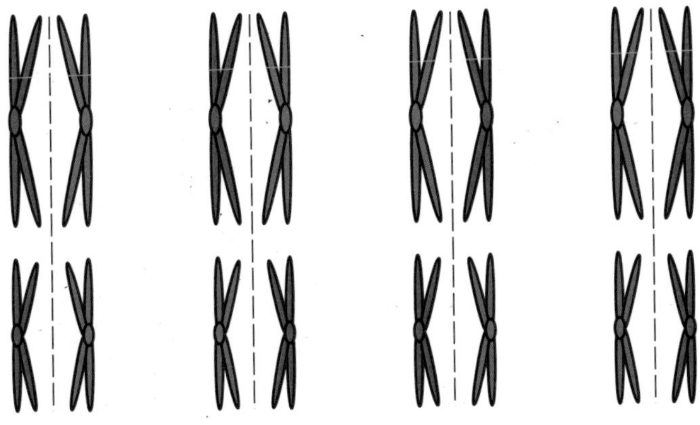

Figure 22.2 The possible combinations of homologous chromosomes created by independent assortment

In a similar way the orientation of the chromatids on the equator in metaphase 2 is random. Therefore the combination of chromatids migrating to each pole is random. This is called independent assortment of sister chromatids.

Exam tip

In a question about creating variation remember that random assortment occurs twice — once in division 1 and again in division 2.

Random fusion of gametes

Fertilisation is the combination of the paternal DNA from a sperm with the maternal DNA in the egg. This is essential to re-establish the diploid number of chromosomes. As a result of meiosis the combination of alleles in every egg cell is slightly different. Equally, the combination of alleles in every sperm cell is different. Any sperm cell can fertilise any egg (random fertilisation) so the potential for increased genetic variation is huge.

Typical mistake

Many students treat fertilisation as if it were part of meiosis. This is not the case. Meiosis is the division of the nucleus and formation of haploid cells. Fertilisation is the combination of two haploid cells to produce a single diploid cell.

Using genetic diagrams

Genetic diagrams are a clear way to set out and explain the results of a cross. There are certain rules you should follow:

- Always use the standard format for diagrams.
- Start with the full **phenotype** and **genotype** of the parents.
- Remember that each gamete contains one gene from each pair of alleles in the parent.
- Draw a circle around the genes in a gamete. This signifies it is a gamete.
- Remember to combine each male gamete with each female gamete. This represents random fertilisation. Using a punnett square will help.
- Write out the full phenotypes of the offspring based on the genotypes produced in the diagram.

The **phenotype** refers to the characteristics expressed in the individual. These characteristics result from the proteins produced by the genes.

The **genotype** is the combination of alleles held in the nucleus.

Exam tip

When drawing a genetic diagram make sure each step is clear. Follow the standard format described and use the same symbols for genes given in the question.

Monogenic inheritance

REVISED

Monogenic inheritance involves one gene. There may be alternative alleles. For example, in *Drosophila* there are two types of wing: long (wild type) and short (vestigial). The wild type is coded for by the **dominant allele** (**L**) and the vestigial is coded for by the **recessive allele** (**l**). A diagram showing a cross between an individual **homozygous** for allele **L** and another individual homozygous for allele **l** would look like Figure 22.3.

Monogenic inheritance involves one gene.

The **dominant allele** is expressed in the phenotype even if a recessive allele is present.

The **recessive allele** is only expressed if no dominant allele is present.

Homozygous means that both the alleles in an individual are the same.

Parental phenotypes	Long wing (wild type)	×	Vestigial wing	
Parental genotypes	LL		ll	
Parental gametes	(L)		(l)	
F_1 *genotype*	Ll			
F_1 *phenotype*	All long wing (wild type)			
F_1 *phenotypes*	Long wing	×	Long wing	
F_1 *genotypes*	Ll		Ll	
F_1 *gametes*	(L) (l)		(L) (l)	

		Male gametes (L)	Male gametes (l)
Female gametes	(L)	LL	Ll
Female gametes	(l)	Ll	ll

F_2 *genotypes*	LL	Ll	Ll	ll
F_2 *phenotypes*	Long wing	Long wing	Long wing	Vestigial wing

F_2 *phenotypic ratio* 3 long wing : 1 vestigial wing

Figure 22.3 A genetic diagram showing monogenic inheritance

All the F_1 individuals are **heterozygous** and express the wild type wing. The F_2 generation shows a phenotypic ratio of 3 wild type to 1 vestigial wing.

Heterozygous means that the two alleles in an individual are different.

Dihybrid inheritance

REVISED

Dihybrid inheritance involves two genes. Again, each gene can have alternative alleles. For example, in *Drosophila* the wild-type body colour is grey (coded for by the allele **E**), but a recessive ebony version exists (coded for by the allele **e**). A cross between a homozygous grey-bodied, long-winged fly and a homozygous ebony-bodied, vestigial-winged fly should be shown as in Figure 22.4.

Parental phenotypes Long wing, grey body (wild type) × Vestigial wing, ebony body
Parental genotypes WWEE wwee
Parental gametes (WE) (we)

F_1 genotype WwEe
F_1 phenotype All long wing, grey body (wild type)

F_1 phenotypes Long wing, grey body × Long wing, grey body
F_1 genotypes WwEe WwEe
F_1 gametes (WE) (We) (wE) (we) (WE) (We) (wE) (we)

		Male gametes			
		(WE)	(We)	(wE)	(we)
Female gametes	(WE)	WWEE	WWEe	WwEE	WwEe
	(We)	WWEe	WWee	WwEe	Wwee
	(wE)	WwEE	WwEe	wwEE	wwEe
	(we)	WwEe	Wwee	wwEe	wwee

F_2 genotypes W_E_ W_ee wwE_ wwee
F_2 phenotypes Long wing grey body; Long wing ebony body; Vestigial wing grey body; Vestigial wing ebony body

F_2 phenotypic ratio 9 long wing grey body : 3 long wing ebony body : 3 vestigial wing grey body : 1 vestigial wing ebony body

Figure 22.4 A genetic diagram showing dihybrid inheritance

The 9:3:3:1 ratio shown in Figure 22.4 is typical of a cross between individuals that are homozygous for both characteristics.

Multiple alleles

REVISED

Multiple-allele inheritance occurs where there are more than two versions (alleles) of a gene. As cells contain two homologous chromosomes they can only contain two copies of each gene. This creates interesting inheritance patterns because different combinations of alleles will be expressed in different ways, producing a range of phenotypes. For example, in humans there are three alleles for blood groups — **A**, **B** and **o**. Alleles **A** and **B** are dominant to allele **o**. A cross between heterozygous individuals showing the **A** and **B** phenotypes would be shown as in Figure 22.5.

Parental phenotypes Group A × Group B

Parental genotypes I^AI^O × I^BI^O

Gametes (I^A) (I^O) × (I^B) (I^O)

		Female gametes	
		I^B	I^O
Male gametes	I^A	I^AI^B	I^AI^O
	I^O	I^BI^O	I^OI^O

F_1 genotypes I^AI^B I^AI^O I^BI^O I^OI^O

F_1 phenotypes (blood groups) AB A B O

Figure 22.5 A genetic diagram showing inheritance of multiple alleles

Sex linkage

REVISED

Examples of sex-linked genes in humans include haemophilia and colour blindness. **Sex linkage** is seen where certain genes are found on only one of the chromosomes that determine sex. This usually means that the Y chromosome is missing the gene and a male (XY) has only one copy of the gene.

In diagrams involving sex linkage the X and Y chromosomes are included with the relevant genes shown as superscripts. Figure 22.6 shows the inheritance pattern found when a homozygous white-eyed female fruit fly is crossed with a red-eyed male. Red eye is the dominant feature (allele **R**); white eye is the recessive feature (**r**). The offspring (F_1) of the original parents are then crossed to produce an F_2 generation.

Sex linkage is where a certain characteristic occurs more often in one gender than in the other. It is caused by presence of the gene on only one sex chromosome.

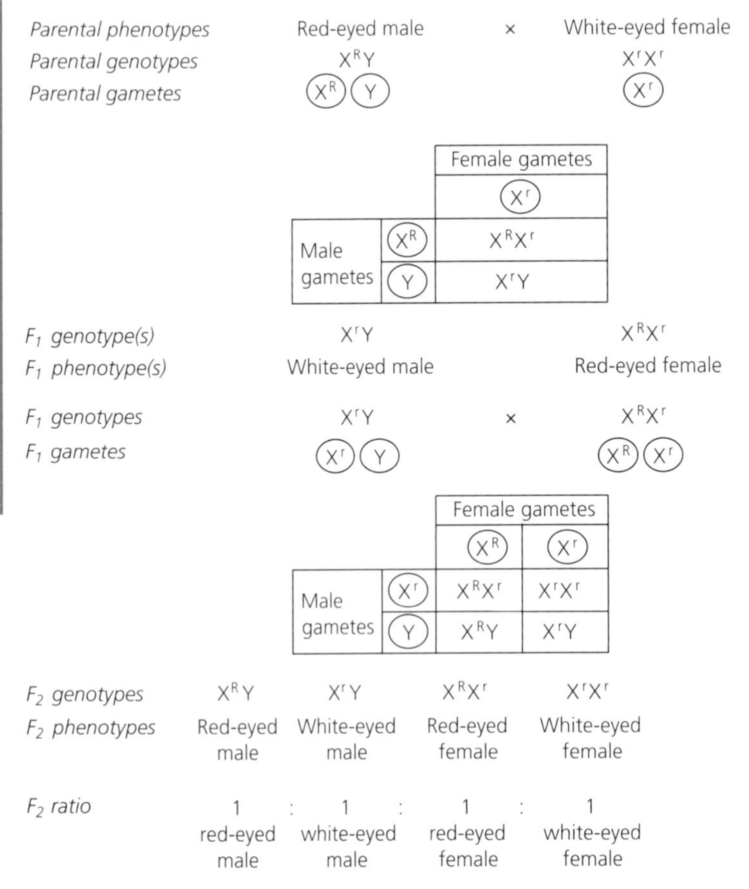

Figure 22.6 A genetic diagram to show inheritance of sex-linked genes

Now test yourself

1 Draw a genetic diagram to show the expected phenotype ratio in the F_2 generation when a pure-breeding (homozygous) red-eyed female fly is crossed with a white-eyed male and the F_1 are crossed.

Answer on p. 230

TESTED

Codominance

REVISED

Codominance occurs when there is no dominant and no recessive characteristic — both alleles contribute to the phenotype. For example, in the snapdragon flower colour can be red (genotype **RR**), white (genotype **WW**) or pink (genotype **RW**). Both alleles **R** and **W** are equally dominant (Figure 22.7).

Codominance is seen where both alleles contribute to the phenotype.

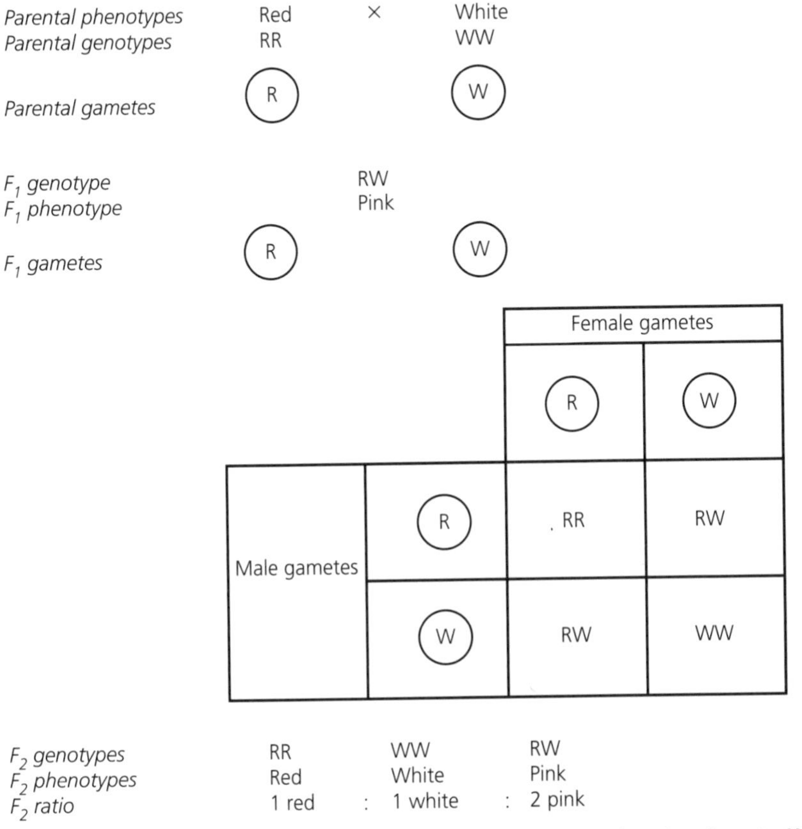

Figure 22.7 A genetic diagram showing inheritance of codominant alleles

Now test yourself

2 Explain the difference between the following pairs of terms:
(a) dominant and codominant
(b) chromosome and chromatid
(c) gene and allele

Answer on p. 230

TESTED

Now test yourself

TESTED

3 Draw a genetic diagram to show why you would not expect to get any white offspring when a roan cow is crossed with a red bull. (Roan cattle have a combination of white and red hairs. Coat colour is a codominant feature.)

Answer on p. 230

Autosomal linkage

REVISED

Autosomal linkage is seen where two separate genes are found on the same autosome (non sex chromosome). As a result they tend to be inherited together and are only separated when a cross-over occurs between the two genes on the chromosome. This means that the offspring usually show the same combination of characteristics as seen in the parent (parental types). Non-parental types or cross-over types are less common.

- With no crossing over the genes stay in their parental combinations and the offspring show a 3:1 ratio of phenotypes, just as expected from a monogenic cross.
- When crossing over occurs the proportion of cross-over types will be low and the actual proportion depends upon how often cross-overs occurred between the two loci on the chromosome.

For example, in a hypothetical cross between one individual homozygous for alleles **A** and **B** and a second individual homozygous for alleles **a** and **b** all the F_1 generation will be heterozygous. When two heterozygous F_1 individuals are crossed the results will differ according to the level of linkage. The possible results of this cross are shown in Table 22.1.

Table 22.1 **The results of a cross involving autosomal linkage compared with an unlinked cross (parental types are shaded green and cross-over types are shaded red)**

Situation	Genotype ratios				Comment
No linkage	1 **AABB** 2 **AABb** 2 **AaBB** 4 **AaBb**	1 **AAbb** 2 **Aabb**	1 **aaBB** 2 **aaBb**	1 **aabb**	Expected 9:3:3:1 ratio of phenotypes seen
Linked and no cross-overs	1 **ABAB** 2 **ABab**			1 **abab**	3:1 ratio of phenotypes.
Linked with some cross-overs	**ABAB** **ABab**	**Abab** **AbAb**	**aBab** **aBaB**	**abab**	The ratios of genotypes and phenotypes are variable There will always be more parental types than cross-over types If the genes are close together on the chromosome there will be fewer cross-over types
Note: where linkage occurs the genotype is shown as **ABab** rather than **AaBb**					

Epistasis

REVISED

Epistasis is when two or more genes (on different loci) interact to influence the phenotype. This is seen where a chain of reactions (a reaction pathway) leads to production of the protein that produces the phenotype. Each step in the pathway requires an enzyme and each enzyme is produced by a different gene.

Epistasis is where genes on different loci interact with each other.

Recessive epistasis

- To make a purple compound both enzymes A and B are needed. Therefore the organism must possess dominant alleles of both genes A and B (Figure 22.8).

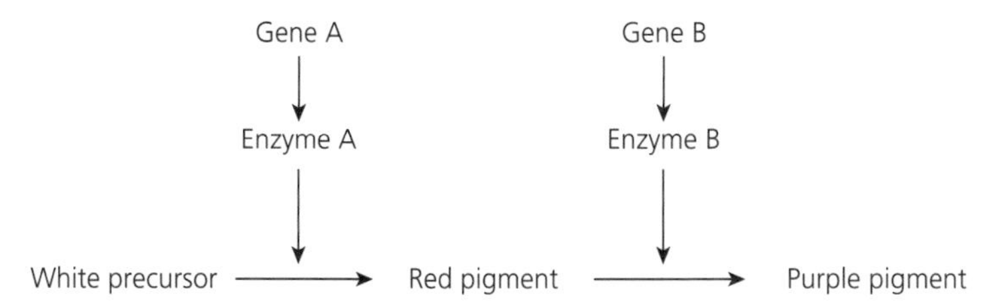

Figure 22.8 Recessive epistasis 1

- If allele **A** is not present then enzyme A is not made and the white precursor is not converted to a red compound. The colour remains white.
- If allele **A** is present but allele **B** is not present then enzyme A is made but enzyme B is not made. The white precursor is converted to a red compound, which cannot be converted to a purple compound.
- If allele **A** is not present then it does not matter if **B** is present or not as there will be no red compound to be converted to purple compound.

This is called recessive epistasis — the presence of recessive alleles for one gene (gene A) affects the expression of another gene (gene B). The ratio of phenotypes observed resulting from a cross will usually be a modified version of the 9:3:3:1 ratio (Table 22.2).

Table 22.2 The modified phenotype ratio resulting from recessive epistasis 1

Normal genotype ratio	1 **AABB** 2 **AABb** 2 **AaBB** 4 **AaBb**	1 **AAbb** 2 **Aabb**	1 **aaBB** 2 **aaBb**	1 **aabb**
Phenotypes	9 purple	3 red	4 white	

In an alternative biochemical pathway the results might be different (Figure 22.9).

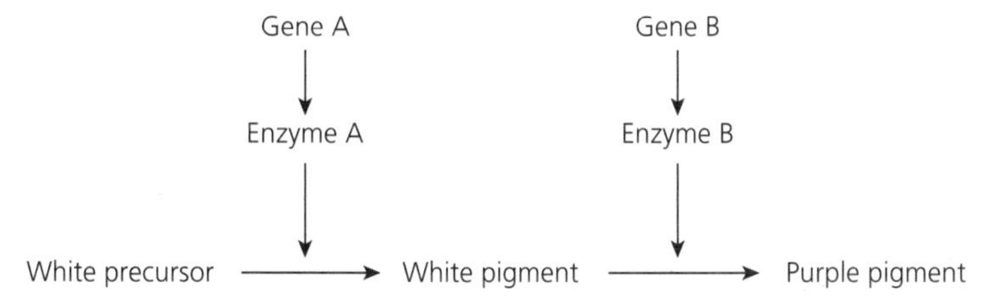

Figure 22.9 Recessive epistasis 2

In this situation gene **A** converts the precursor to a white pigment, so the presence of allele **A** does not change the visible phenotype. But if allele **A** is not present it will affect the expression of gene **B** (Table 22.3).

Table 22.3 The modified phenotype ratio resulting from recessive epistasis 2

Normal genotype ratio	1 **AABB** 2 **AABb** 2 **AaBB** 4 **AaBb**	1 **AAbb** 2 **Aabb**	1 **aaBB** 2 **aaBb**	1 **aabb**
Phenotypes	9 purple	7 white		

Dominant epistasis

- To make a purple compound enzyme A is needed from gene A — the dominant allele **A** must be present (Figure 22.10).
- However, if the dominant allele of gene B is present it produces enzyme B. Enzyme B competes with enzyme A for the substrate. If enzyme A is the better competitor then the precursor is converted to purple rather than to red.
- If allele **A** is not present then enzymes produced by gene B will be active — these convert the precursor to a red pigment. If neither allele **A** nor **B** is present then no enzymes are made and the precursor remains white.

This is called dominant epistasis because the presence of the dominant allele of a gene (gene A) affects the expression of another gene (gene B).

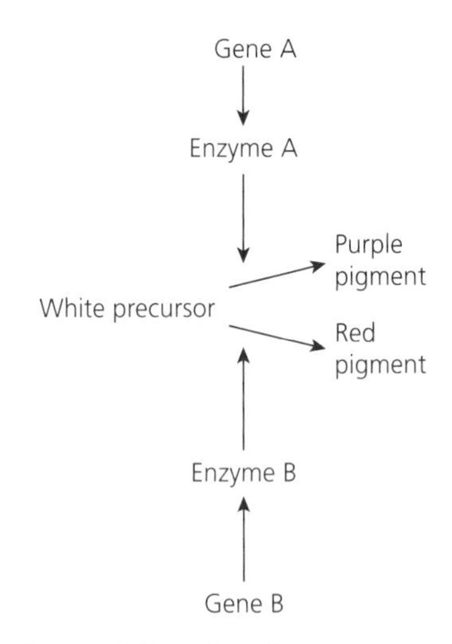

Figure 22.10 Dominant epistasis 1

Table 22.4 The modified phenotype ratio resulting from dominant epistasis 1

Normal genotype ratio	1 **AABB** 2 **AABb** 2 **AaBB** 4 **AaBb**	1 **AAbb** 2 **Aabb**	1 **aaBB** 2 **aaBb**	1 **aabb**
Phenotypes	12 purple	3 red		1 white

In an alternative biochemical pathway the results may be different (Figure 22.11).

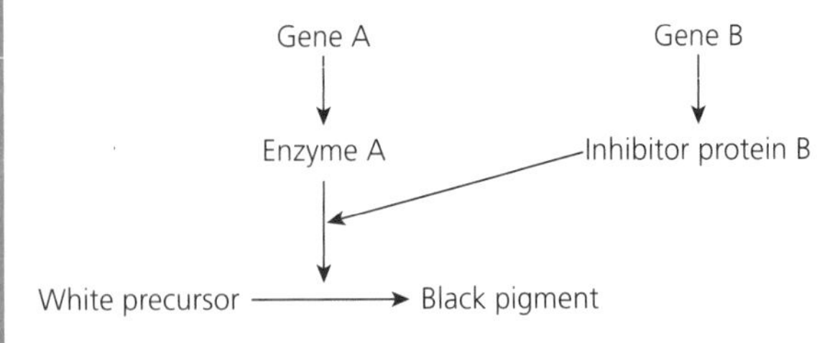

Figure 22.11 Dominant epistasis 2

Gene A produces an enzyme that converts the white precursor to a black pigment. If allele **A** is present then the enzyme is produced. However, gene B produces a protein that inhibits the enzyme. Therefore, if allele **B** is present the action of gene A is inhibited and the white precursor is not converted to a black pigment (Table 22.5).

Table 22.5 The modified phenotype ratio resulting from dominant epistasis 2

Normal genotype ratio	1 **AABB** 2 **AABb** 2 **AaBB** 4 **AaBb**	1 **aaBB** 2 **aaBb**	1 **aabb** 1 **AAbb**	2 **Aabb**
Phenotypes	13 white			3 black

Exam tip

You are not expected to draw genetic diagrams for examples of epistasis or autosomal linkage. However, you may be asked to interpret genetic diagrams of epistasis or autosomal linkage.

Typical mistake

Many students confuse dominant and recessive epistasis. Unless you are asked to state whether it is dominant or recessive, simply call any gene interaction epistasis without giving further detail.

Chi-squared testing

The chi-squared test is a statistical test used to determine whether the observed results of a cross (*O*) differ significantly from the expected results (*E*). The expected results are theoretical results usually calculated using a genetic diagram. The formula is:

$$\text{chi-squared } (\chi^2) = \Sigma \frac{(O - E)^2}{E}$$

This formula calculates a value for the difference between *O* and *E* that can then be used to determine whether the difference is due to chance. The larger the value of χ^2 calculated the greater the difference between the observed and expected results and therefore the lower the probability that the observed results were created by chance.

The test involves making a null hypothesis. This could simply be 'there is no difference between the observed and expected results'.

Once the value of χ^2 is calculated using the formula it is compared with a critical value of χ^2, which can be looked up on a table of probabilities for χ^2. If the calculated value of χ^2 is lower than the critical value then we can accept our null hypothesis as the differences could arise due to chance. However, if the calculated value of χ^2 is greater than the critical value then there is a low probability that the differences could arise by chance. The observed results are significantly different from the expected.

Remember that the critical value of χ^2 is agreed to be at $p = 0.05$ and the number of degrees of freedom in your cross is one less than the number of categories in the results.

Exam tip

You are unlikely to be asked to carry out a complete calculation in an examination. You may need to carry out part of the calculation, but it is more likely that you would be asked to discuss the use of the test or to interpret the results of a test.

Variation

Continuous variation

REVISED

Continuous variation is seen where there are no distinct groups or categories and there is a full range between two extremes; for example, height and body weight. The continuously variable feature can be quantified and data are usually presented in the form of a histogram. This form of variation is caused by:

- a number of genes interacting together
- the environment

Continuous variation is variation that shows a complete range between two extremes, with no distinct groups.

Discontinuous variation

REVISED

Discontinuous variation is seen where there are distinct groups or categories, with no in-between types; for example, gender and possession of resistance or immunity. The discontinuously variable feature cannot be quantified — it is qualitative. Data are usually presented in the form of a bar chart.

This type of variation is usually caused by one gene, with the presence of a dominant allele producing one phenotype, while the absence of that dominant allele in the double recessive is expressed as a different phenotype.

Discontinuous variation is variation that produces distinct groups.

Typical mistake

Students tend to plot bar charts with no gaps between the bars. There should be a gap between the bars and the bars should be the same width.

The need for variation

REVISED

Variation is an essential part of evolution. Evolution is achieved by natural selection. This is the selection of the best-adapted individuals, allowing them to survive and pass on their alleles to the next generation. If there were no variation between individuals, then none would be better adapted than any others. Selection could not occur.

Revision activity

Draw a mind map with 'variation' at the centre. Leave space to continue expanding the map later.

Factors that affect evolution

A selective force is some aspect of the environment that places a **selection pressure** on the species. It makes it more difficult for individuals to survive and reproduce. Therefore, those individuals with better adaptations survive more easily and are more likely to reproduce. This allows them to pass on their alleles to the next generation.

Selection pressure is a force that gives a selective advantage to one phenotype over another.

We can view all the alleles in a population as a gene pool. Each individual contributes two alleles to the gene pool. If many individuals possess one particular allele then that allele makes up a high proportion of the gene pool — it has a high frequency. If the frequency of alleles in the gene pool remains constant then the population does not change. However, if the frequency of alleles changes then the population changes and evolution is occurring.

Selective forces are forces that can change the frequency of alleles by placing selective pressure on the individuals.

Stabilising selection

REVISED

Assuming that a species is well adapted to the environment and the environment stays constant there will be no evolutionary change. This is because any change away from the well-adapted form is unlikely to be successful and will not pass on more alleles to the next generation.

Directional selection

Evolution occurs when the environment changes and places a selection pressure on the species. Genetic variation is caused by the presence of different alleles in the population. If the selection pressure favours those individuals that possess a particular allele, then those individuals are placed at a selective advantage. They will reproduce more frequently and a higher proportion of the next generation will possess that allele. The allele frequency changes.

One example of directional selection is the change from light to dark colour that was observed in the peppered moth (*Biston betularia*) during the industrial revolution.

Exam tip

Always refer to changes in allele frequency rather than gene frequency.

Genetic drift

REVISED

As individuals are born or die there may be small changes in the frequency of each allele. In a large population this will have little or no effect. However, in a small population the loss of one or two individuals could have a big effect on the gene pool — it may even cause the loss of a particular allele from the gene pool. Alternatively, the survival of one or two unusual individuals could increase the frequency of a certain allele. These changes are known as **genetic drift**.

Genetic drift is when allele frequencies change by chance.

Genetic bottlenecks

REVISED

A genetic bottleneck occurs when the population reduces in size and then starts to increase in size again. The new population contains only the alleles that were in the smallest population and not those alleles that may have existed before. This means that the new increasing population has a lower genetic diversity and lower phenotypic diversity.

Founder effect

REVISED

The founder effect is seen where an isolated population grows from a small sample of the entire population. The small number of individuals founding the new population may not have a representative proportion of all the alleles in the main population. As a result the new population may lack genetic diversity.

The Hardy-Weinberg principle

The Hardy-Weinberg principle is used to calculate the frequencies of alleles in a population. A population will contain individuals of three genotypes: **AA**, **Aa** and **aa**, where **A** is the dominant allele, and **a** is the recessive allele. Two equations are used:

$p + q = 1$ (Equation 1)

$p^2 + 2pq + q^2 = 1$ (Equation 2)

where p is the frequency of the dominant allele and q is the frequency of the recessive allele. Therefore, p^2 is the frequency of homozygous dominant individuals (**AA**), $2pq$ is the frequency of heterozygous individuals (**Aa**) and q^2 is the frequency of homozygous recessive individuals (**aa**).

The individuals with the homozygous recessive genotype can be recognised — they show the recessive phenotype. In a population where 9% of individuals show the recessive feature, this is a frequency of 0.09. Therefore, $q^2 = 0.09$. So $q = \sqrt{0.09} = 0.3$.

To find the value of p we enter $q = 0.3$ into equation 1: $p + 0.3 = 1$. Therefore, $p = 0.7$.

Knowing the values of p and q we can calculate the frequencies of the different genotypes in the population:

frequency of the homozygous dominant, $p^2 = 0.7^2 = 0.49$

frequency of the heterozygous genotype, $2pq = 2 \times 0.7 \times 0.3 = 0.42$

This means that our population consists of 9% homozygous recessive individuals, 49% homozygous dominant individuals and 42% heterozygous individuals.

Now test yourself

4 Calculate the frequency of heterozygous individuals in a population where 84% of the population show the dominant characteristic.

Answer on p. 230

TESTED

Assumptions

REVISED

The Hardy-Weinberg principle does assume that:

- The population is large.
- Mating is random.
- There is no mutation, immigration or emigration.
- No selective pressure is operating.

Isolating mechanisms

The evolution of a new species occurs when members of the same species are unable to freely interbreed. This is called reproductive isolation. When this occurs any genetic changes do not spread throughout the species and one population might accumulate changes not seen in another population. If enough changes occur such that the members of one population can no longer breed successfully (producing fertile offspring) with the members of another population then speciation has occurred.

There are two isolating mechanisms:

- Geographical barriers, such as large rivers, seas or mountains, prevent the populations mixing. This is known as allopatric speciation.
- Reproductive barriers, such as differences in courtship rituals, slight biochemical changes or variation in timing of the breeding season may prevent reproduction between individuals in the same place — this is called sympatric speciation.

Natural and artificial selection

Natural selection and artificial selection operate in the same way — a selective pressure causes allele frequencies to change.

In natural selection the selective forces are biotic or abiotic aspects of the environment. An allele that confers a selective advantage to the individual is favoured because that individual is more likely to survive and breed to pass that allele on to the next generation. Natural selection affects all the features of an organism and will proceed slowly unless the environment changes dramatically.

In artificial selection the selective force is human choice. An individual that shows a desired characteristic is selected to breed. This means that the allele causing that desired characteristic is passed on to the next generation. Artificial selection affects only the one or two characteristics chosen by the human and can proceed quickly because selection pressure is very strong.

Artificial selection of modern dairy cows

REVISED

Selection in dairy cows can take a long time because cows take many months or years to reach their full milk-producing potential. Milk yield shows continuous variation. Females are selected by their performance — a high milk yield is desirable. Males are selected by testing the performance of their female offspring. Once a suitable bull has been identified his sperm can be collected and stored — possibly frozen for many years. The sperm can be delivered by artificial insemination to a large number of suitable females. Again the offspring can be tested and, over a number of generations, the milk yield per cow can be increased.

Typical mistake

Many students omit to state that selection must continue over several generations.

Artificial selection of bread wheat

REVISED

Modern bread wheat (*Triticum aestivum*) is the result of artificial selection over thousands of years:

- Einkorn wheat (*Triticum urartu*) crossed naturally with wild grass (*Aegilops* sp.) to produce a hybrid. This hybrid was sterile but became fertile after a chromosome mutation. This was wild emmer wheat (*Triticum turgidum dicoccoides*).
- Selection by farmers in the Middle East 10 000 years ago resulted in the wild emmer wheat giving rise to cultivated emmer wheat (*Triticum turgidum dicccum*).
- This cultivated emmer wheat was bred with other species (*Aegilops tauschii*) to produce a number of hybrids. These hybrids were sterile because they possessed an uneven number of chromosomes.
- Chromosome mutation of the sterile hybrid led to a doubling in the number of chromosomes, making the hybrid fertile. This fertile hybrid is known as spelt wheat (*Triticum aestivum spelta*).
- Further gene mutation and selection has led to modern bread wheat (*Tritcum aestivum aestivum*).

More recently, plant breeders have improved the species by adding genes from other varieties and species. These have included genes to provide resistance to disease such as stem rust and other fungal attack. Introduction of such genes involves crossing the high-yield variety with another variety that shows the desired characteristic. The offspring are then tested to check that they show the desired combination of characteristics. If the desired combination is shown then the offspring are back-crossed with the high-yield variety for several generations to ensure that the offspring are pure breeding.

Exam tip

You are unlikely to be asked to remember all the names of the different species involved in the evolution of modern bread wheat. They are supplied so that you are familiar with the names should they be given as part of a question.

The need for conservation

REVISED

The introduction of new characteristics through selective breeding highlights the need to maintain the genetic diversity of species, including conserving rare breeds and wild-type plants. Rare breeds and wild-type plants possess combinations of genes and alleles that enable them to survive in a range of environmental conditions. If these combinations are lost because we allow rare breeds and wild-type plants to become extinct then we will no longer have that genetic resource to exploit.

Ethical considerations

REVISED

Artificial selection is used to alter the phenotype to achieve something desired by humans. It often alters one characteristic with little concern about other characteristics of the organisms. This can lead to breeds that would not survive without human intervention.

- Dogs have been domesticated and bred for thousands of years. All dog breeds have evolved from the same origin. Many breeds have become susceptible to certain diseases or conditions as a result of continual inbreeding, for example, labradors are associated with a congenital hip disorder, while bulldogs have difficulty giving birth.
- Many domesticated animals retain characteristics that make them friendly and docile but unable to defend themselves against predators.
- Animals bred to have lean meat with little fat may not be sufficiently well insulated to survive in cold temperatures.

Revision activity

Expand your mind map of variation (started earlier in this chapter) by including ideas about selection (natural and artificial) and how evolution occurs.

Exam practice

1 (a) Explain what is meant by the term *allele*.. [2]
(b) (i) State the name given to the point at which chromatids break and rejoin. [1]
(ii) Identify the stage of meiosis at which chromatids break and rejoin. [1]
(iii) When the chromatids break and rejoin genetic material can be exchanged. Describe the benefit to the species of this process. [4]

2 The figure shows a pair of homologous chromosomes at the start of meiosis. Two genes and their alleles are shown.

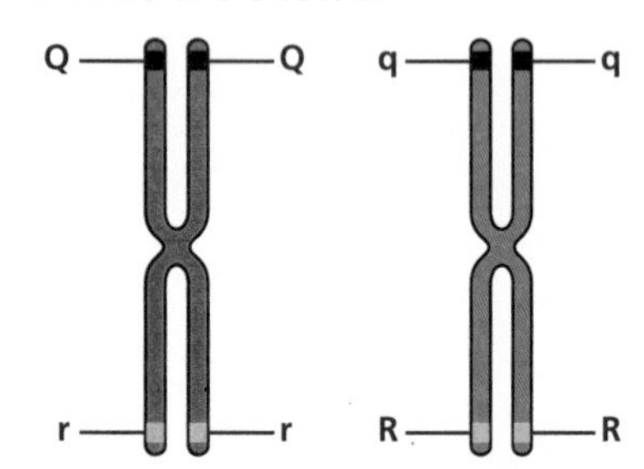

(a) What is the genotype of this individual? [2]
(b) Write down the possible gametes that could be produced assuming that:
(i) there are no cross-overs [2]
(ii) there is one cross-over between the loci for genes *q* and *r* [4]
(iii) there are two cross-overs between the loci for genes *q* and *r* [2]

3 The diagram shows the life cycle of a mammal.

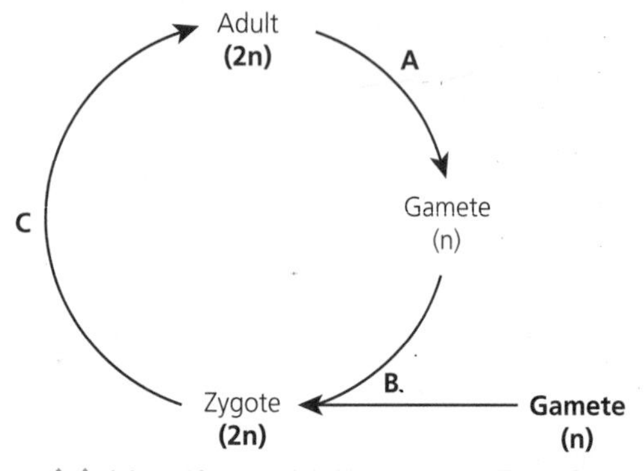

(a) Identify at which stage, A, B or C, the following processes take place:
(i) meiosis [1]
(ii) mitosis [1]
(iii) fertilisation [1]
(b) Explain why the gametes contain the haploid (*n*) number of chromosomes. [3]

4 (a) Explain what is meant by directional selection. [2]
(b) Explain why variation is essential for selection to occur. [2]
(c) Using one example, describe how artificial selection can be achieved. [5]

Answers and quick quiz 22 online

ONLINE

Summary

By the end of this chapter you should be able to:

- describe the contribution of genetic and environmental factors to phenotypic variation
- use genetic diagrams to show patterns of inheritance in crosses involving monogenic and dihybrid inheritance, multiple alleles, sex-linkage and codominance
- explain autosomal linkage and epistasis using phenotypic ratios to identify linkage
- use the chi-squared test
- distinguish between continuous and discontinuous variation
- explain the factors that affect evolution of a species, including stabilising selection, directional selection, genetic drift, the bottleneck effect and founder effect
- use the Hardy-Weinberg principle
- explain the role of isolating mechanisms (allopatric and sympatric)
- discuss artificial selection, its use in selective breeding and ethical considerations

23 Manipulating genomes

Sequencing genomes

A genome is all the genetic information inside a cell or organism. The genome includes coding DNA (consisting of genes that code for the structure of polypeptides) and non-coding DNA (which carries out a range of regulatory functions).

Sequencing a genome involves a series of steps:

- DNA is extracted from from the cells.
- The DNA is cut into sections of about 100 000 nucleotides in length using restriction endonuclease enzymes.
- These sections are placed into bacterial artificial chromosomes (BACs) and transferred to cells of *E. coli*.
- These *E. coli* cells are grown in culture to clone many copies of the DNA.
- The DNA is extracted from the *E. coli*.
- The DNA is cut into shorter fragments (of up to 1000 nucleotides) using a range of different restriction enzymes. This ensures that the sections are cut in a variety of different places so that the fragments overlap rather than producing many fragments of the same length.
- The fragments are separated by gel electrophoresis.
- Each fragment is sequenced and the overlapping fragments are compared to reassemble the whole BAC sequence.

Exam tip

You are only expected to know an outline of this process; there is no need to confuse it with too much detail.

Now test yourself TESTED

1 Explain why it is essential to use a range of different restriction endonucleases.

Answer on p. 230

Revision activity

Draw a flow chart to represent the sequencing process.

Sanger sequencing

REVISED

Sanger sequencing involved using a single-stranded length of DNA to act as a template. A complementary strand was allowed to grow on the template in a solution that contained all four nucleotides and the enzyme DNA polymerase. A modified version of one nucleotide was added to the solution — this was labelled and also stopped growth of the complementary strand. Sanger produced a series of lengths of single-stranded DNA, each ending in that radioactively labelled nucleotide. The process was repeated in separate vessels for each of the four bases. The lengths of single-stranded DNA could then be separated by gel electrophoresis (Figure 23.1) in four separate wells on the same gel plate.

Sanger's method has been automated using nucleotides modified by attachment of a fluorescent probe which stops growth of the complementary strand. Each base has a different colour probe. As the mixture is separated by electrophoresis in a capillary tube the shortest lengths move most quickly and are detected as they move past a sensor, followed by fragments of increasing length. The detector reads the coloured probe on each fragment and produces a sequence.

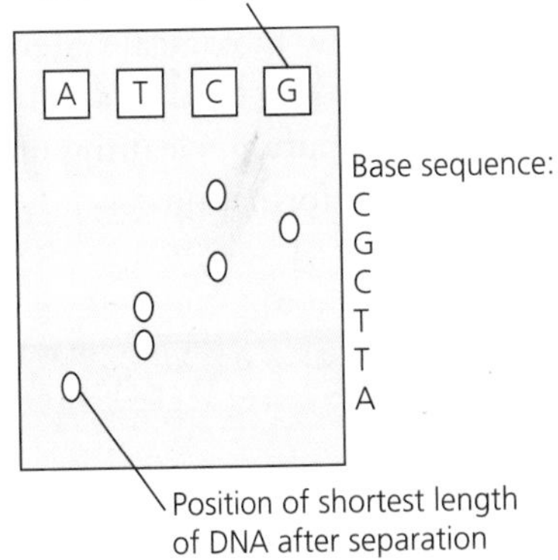

Figure 23.1 Sequencing by separation on a gel electrophoresis plate

High-throughput sequencing

REVISED

In the late 1990s and in the early part of this century a number of new techniques were devised to speed up the process of sequencing. Many of these were based on reading the bases as they were combined with a single-stranded DNA template rather than reading from the end of a sequence as in the Sanger process.

Pyrosequencing is one example of these new techniques. This uses the enzyme luciferase to generate photons of light as nucleotides are added to the sequence. Individual DNA molecules are grown in separate water droplets in an oil/water emulsion. The technique can read millions of data points simultaneously and produces huge amounts of information that must be analysed by special computer programmes.

Using genome sequences

REVISED

The sequence of nucleotides in the genome or in a particular gene can be compared with the sequence in another individual or species. A new branch of biology known as bioinformatics has developed to store and analyse the huge amounts of data generated. Bioinformatics relies on computers, microchips and specialist software packages. This can be useful to:

- ascertain how important the gene is
- work out evolutionary relationships — the more similar the sequence the more closely related the species
- compare the genome of a pathogenic organism with that of a similar, non-pathogenic organism. This enables us to find out what genes or base sequences are important in causing the disease. This could help to design drugs to combat the disease.
- identify which genes cause inherited diseases by comparing the base sequence from a healthy person with that from someone who has an inherited disease
- investigate the effects of a specific mutation by manufacturing the mutation and inserting it into a microorganism such as yeast
- investigate the relationship between genotype and phenotype
- enable scientists to predict the amino acid sequence in a protein from the base sequence in the gene

Synthetic biology

The DNA sequences provided by genome sequencing can be used as building blocks for new biological systems. Synthetic biology involves designing and building useful biological devices and systems. The aim is to build new biological systems that can fulfill certain requirements, such as providing food, diagnosing illness, producing medication, detecting pollution, cleaning up pollution and even storing or processing information.

Now test yourself

TESTED

2 Explain how a genome sequence can be used to work out evolutionary relationships.

Answer on p. 230

Revision activity

Draw a mind map with 'genome sequence' at the centre. Don't forget to include DNA structure, the genetic code and protein synthesis as synoptic topics.

DNA profiling

DNA profiling is a way of identifying DNA. It is carried out using non-coding DNA, known as short tandem repeat sequences. This DNA consists of short sequences of DNA that are repeated many times. These sequences appear at a number of sites in the genome and the number of repeats at each site is variable. As a result, each individual has a unique pattern of short tandem repeats.

The DNA is cut using restriction enzymes that cut the DNA at a specific base sequence. The fragments vary in size from individual to individual. They are separated by gel electrophoresis. The fragments are then stained and a banding pattern is produced. This banding pattern is specific to the individual from whom the DNA came.

Applications of DNA profiling

REVISED

DNA profiling has many uses in forensic science. It can:

- identify people involved in criminal activity by comparing DNA found at the scene with that of suspects
- prove someone's innocence
- identify body parts or remains

Another use of DNA profiling is in the analysis of risk of disease:

- Presymptomatic testing can be used for people who have a family member with a genetic disorder but who have no symptoms themselves, e.g. testing for Huntingdon disease.
- Predictive testing can identify alleles or combinations of alleles that increase a person's chances of developing a disorder such as certain types of cancer or heart disease.

DNA profiling can also be used to establish maternity or paternity when this is disputed.

The polymerase chain reaction (PCR)

Sequencing DNA requires a large number of molecules. If only a small sample is available it can be amplified (replicated many times). This is achieved by the **polymerase chain reaction**. The sequence of steps in PCR is as follows:

- Mix the sample of DNA with extra DNA nucleotides and DNA polymerase.
- Heat to 95°C, which breaks the hydrogen bonds holding the two strands together and splits the double helix to form single strands with exposed bases.
- Cool to 55°C and add short lengths of single-stranded DNA known as primers, which bind to the strands to be replicated — this produces double-stranded sections, which allows the DNA polymerase to bind.
- Raise the temperature to 72°C, which is the optimum temperature for this DNA polymerase (taq polymerase). The DNA polymerase adds nucleotides to the double-stranded section until complete new double-stranded molecules are created.
- Repeat the cycle many times.

The **polymerase chain reaction** is a chemical reaction that replicates DNA.

Now test yourself

3 What is special about taq polymerase?

Answer on p. 230

TESTED

Electrophoresis

Electrophoresis is a technique used to separate pieces of DNA. It is similar to chromatography but a gel substrate is used in place of paper or silica. The gel contains wells into which the sample of DNA is placed and is covered by a buffer solution to prevent it drying out. The DNA fragments are free to diffuse through the gel but diffuse slowly due to their size. They can be made to move more quickly in one direction by application of a potential gradient.

DNA is negatively charged and so the fragments move towards the positive electrode end of the gel. The shorter lengths of DNA are smaller and travel more quickly. Therefore the fragments of different lengths are separated.

Electrophoresis is a technique used to separate lengths of DNA according to their size.

DNA probes

REVISED

The lengths of DNA separated by electrophoresis are not visible. They need to be labelled in some way so that the desired gene or nucleotide sequence can be identified. The DNA can be labelled using a DNA probe or, more specifically, a **gene probe**. A DNA probe:

- is a short length of DNA
- is single stranded
- has a specific sequence that complements a sequence in the desired fragment of DNA (possibly a whole gene)
- is labelled in some way using:
  - nucleotides with ^{32}P, which shows up on a photographic plate
  - a specific stain
  - a fluorescent molecule that glows in UV light

A **gene probe** is a short length of DNA used to identify genes with a specific nucleotide sequence.

Separating proteins

REVISED

Proteins can be separated for analysis or for purification. Proteins can be separated in a similar way to DNA but proteins often have surface charges that interfere with their movement. The different shapes of proteins will also affect their movement through a gel. Proteins can be separated by placing them in a well in the centre of the electrophoresis plate so that proteins with different charges can move in opposite directions.

Protein electrophoresis can separate different types of haemoglobin for diagnosis of sickle cell anaemia, leukaemia and thalassaemia.

Genetic engineering

Genetic engineering is also known as recombinant gene technology. **Recombinant DNA** is DNA from one species that has been modified by incorporating one or more genes from another organism — often from another species. Recombinant DNA is produced by genetic engineering.

There are a number of techniques involved:

- extraction of the desired gene
- isolating the gene from other DNA using electrophoresis and probes
- making multiple copies of the gene by the polymerase chain reaction
- placing the gene in a vector
- using the vector to place the gene into the recipient

The recipient then uses the gene to produce the desired product or characteristic.

Recombinant DNA is DNA from one organism that contains a gene from another organism or species.

Typical mistake

Remember the difference between the gene and the gene product — many students confuse the two.

Extraction of genes

REVISED

Restriction endonuclease enzymes are specific to DNA. There is a wide range of endonucleases, each with a different active site that is specific to a particular sequence of nucleotides (the restriction site). Each specific enzyme always cuts DNA at the same sequence of nucleotides. The enzyme hydrolyses the bonds along the sugar phosphate backbone. The DNA is cut in a staggered fashion, leaving a sequence of a few nucleotides unpaired and exposed. This is called a sticky end because it can be used to join to another length of DNA with the complementary sequence of exposed nucleotides.

An alternative way to get the gene is to extract mRNA from cells and use it to make the gene using the enzyme reverse transcriptase. This was the technique used to manufacture the gene for human insulin.

A **restriction endonuclease** is an enzyme that cuts DNA.

Typical mistake

Take care with your wording here. Many students say 'endonucleases cut the gene'. This is not what is required — we want to cut the gene out of the chromosome.

Isolation of the correct gene and amplification

REVISED

Electrophoresis is the technique used to separate pieces of DNA by their length (described above). Once separated the required gene is identified using a suitable gene probe (described above).

Once the gene has been isolated it can be amplified — this means increasing the number of genes available through the polymerase chain reaction (described above).

Using vectors

REVISED

Once a gene or length of DNA has been isolated it must be inserted into the recipient cell. This is done using a **vector**. The gene must be inserted into the vector and then the vector must enter the recipient cell.

Bacteria contain short, circular sequences of DNA called **plasmids**. Plasmids can be used as vectors. They can be modified or created and placed into bacteria in the following way:

- The gene is isolated using a restriction endonuclease, which cuts the gene out of the DNA so that it has sticky ends. A plasmid is cut using the same restriction endonuclease so that it has the same sticky ends as the gene.
- Mixing the isolated gene with cut plasmids allows some plasmids to take up the gene and join up to reform a full circle. This is called annealing. DNA ligase enzymes reform the sugar phosphate backbone of the DNA.
- Bacteria growing in a culture medium containing the newly formed plasmids take up the plasmids (only around 0.25% of the bacterial cells will absorb plasmids). They can be encouraged to take up the plasmid by a number of methods:
  - heat shock — the culture is cooled to 0°C and then quickly heated to 40°C
  - electroporation — a high-voltage pulse applied to the membrane causes temporary gaps to form
  - electrofusion — an electric field around the recipient cells helps the DNA to enter the cells
- Those bacteria that have taken up the plasmids are called transgenic bacteria. They will express the gene in the plasmid and will manufacture the desired product.

A **vector** is a means of inserting DNA into a cell.

A **plasmid** is a short, circular piece of DNA.

Exam tip

Here the endonucleases are used to cut the plasmid open, not to cut bits out of it.

Now test yourself

4 Explain why sticky ends are essential.

Answer on p. 230

TESTED

Why do bacteria take up plasmids?

Bacteria and other microorganisms often reproduce asexually. This means that there is no genetic variation introduced by meiosis and random fertilisation. Taking up DNA from their surroundings introduces genetic variation and increases diversity. This might enable the cell to survive in its environment or to take advantage of certain conditions and thrive where others struggle to survive. Selection and evolution become possible.

Other vectors

- Viruses or bacteriophages can also be used as vectors. DNA placed into a virus is inserted into the recipient cell when the virus attacks the cell.
- Ti plasmids can be inserted into bacteria (*Agrobacterium tumefaciens*), which then infect plants. The bacterium inserts the plasmid into the plant DNA.
- BAC (bacterial artificial chromosomes) are circular lengths of DNA used to clone genes inside bacteria.
- YAC (yeast artificial chromosomes) do the same in yeast.
- Liposomes are tiny balls of fatty molecules that can fuse to cell surface membranes or even cross lipid bilayers. The gene is held inside the ball of fatty molecules and enters the cell as fusion occurs.

Producing human insulin — an example of genetic engineering

REVISED

Human insulin can be produced by engineering bacteria to contain the gene and express it. The human insulin gene is isolated by extracting mRNA from a human pancreas and manufacturing the gene using reverse transcriptase. After amplification the gene is inserted into a bacterial plasmid using DNA ligase enzymes. The bacteria are grown on nutrient agar and allowed to take up the plasmids. The bacteria that have absorbed the plasmids are identified by the use of marker genes and isolated. These bacteria are grown in a fermenter to manufacture human insulin.

Genetic markers

REVISED

Only about 0.25% of cultured bacteria take up the plasmid containing the desired gene. It is important to isolate those that have the plasmid from those that have not because it is pointless to culture the bacteria that will not be able to express the gene and produce the required product. This is achieved using **genetic markers**.

A **genetic marker** is a sequence of DNA used to test that the required gene has been inserted successfully.

The gene is not introduced into the plasmid on its own. The length of DNA inserted into plasmids also contains genes for resistance to two antibiotics (tetracycline and ampicillin). The desired gene is inserted into the middle of the gene for resistance to tetracycline. Therefore bacteria that have taken up the plasmid successfully will be resistant to ampicillin but not to tetracycline. The bacteria are tested to see if they are resistant to the two antibiotics by replica plating:

- Grow the bacteria on normal nutrient agar.
- Blot with sterile velvet to pick up cells from each colony and transfer to agar containing ampicillin — only the bacteria with the plasmid containing the ampicillin resistance gene will grow.

- Blot again and transfer cells to agar containing tetracycline — only the cells with a complete tetracycline resistance gene will grow.
- The cells that grew on the ampicillin but not on the tetracycline are the ones that contain the required gene.

Revision activity

Draw a mind map with 'genetic engineering' at the centre. Include each of the techniques covered above to show the links between them.

Now test yourself

TESTED

5 Explain why the bacteria that have successfully taken up the plasmid will be resistant to one antibiotic but not the second.

Answer on p. 230

Ethical concerns

Genetic engineering is a powerful new technology with huge potential to improve food production and medicine and to reduce human suffering from disease and starvation. However, genetic manipulation of organisms raises many potential ethical concerns (Table 23.1).

Table 23.1

Potential benefits	Concerns
Insulin produced by bacteria can be used to treat all diabetics; if we do not use genetically modified bacteria some people will be left untreated	The possible spread of antibiotic-resistance genes used as markers to 'wild' bacteria
Plants genetically modified to produce an insect toxin can be grown without having to spray insecticides that could harm non-pest insects	Insects could develop resistance to the toxins produced by the plant
Soya has been modified to contain a gene for resistance to certain herbicides; this allows farmers to control weeds without damaging the crop	The herbicide resistance gene could spread to other plants, to create 'superweeds'
Animals including mammals can be 'pharmed' — genetically modified to manufacture pharmaceuticals	Suffering may be caused to farm animals if the genes introduced adversely affect their metabolism
Pathogens such as viruses can be modified and used in the manufacture of antibodies and research	Viruses used as vectors can increase the risk of cancer
Increased yields are possible from genetically modified crops	Possible presence of toxins or allergens in GM crops The reduction of genetic diversity if only certain GM crops are grown This is a new technology, so we simply do not know what the potential risks are

Other issues relate to the companies that carry out the research and produce genetically modified products:

- These are usually commercial companies wanting to make a profit.
- They patent the genetically modified products so that other companies cannot copy them.
- They may be unwilling to share the technology and information.
- The products are often expensive and may not be affordable for poor farmers in less developed countries.

Typical mistake

Some students get into long, heart-felt arguments about the rights and wrongs of genetic engineering and 'playing God'. This is not appropriate at A2.

Gene therapy

Gene therapy does not involve replacing genes. Instead genes are added to the genome of a cell. Most therapies involve using a modified virus to deliver a working copy of a gene (an allele) to patients who lack a working version. When a functioning allele is inserted into a cell the new allele can be expressed, allowing the cell to produce functioning proteins.

Gene therapy involves treating genetic disorders by inserting new alleles.

The potential for gene therapy is huge. Cystic fibrosis has been treated using genes introduced into cells in the airways, using liposomes as vectors. HIV/AIDS patients have been treated using genes that produce molecules that inhibit virus activity. Thalassaemia has also been treated.

Other therapies being developed include those for prostate cancer, metastatic melanoma, neurodegeneration and bladder cancer.

Somatic cell gene therapy

REVISED

Adding genes into a body cell so that the cell can produce a specific protein allows individual cells or tissues in an organism to be augmented. Cells can also be killed in this way — by inserting a gene that enables them to manufacture a foreign antigen, the cells are recognised as foreign and so will be attacked by the immune system. This can be useful for treating cancer.

Treating somatic cells (body cells) is short term, however, and only affects the cells that are treated directly.

Germline gene therapy

REVISED

Treating a fertilised egg means that all cells in that organism will possess the required gene and it will also be passed on to its offspring.

Now test yourself

TESTED

6 Explain why there are more concerns about germline gene therapy than about somatic cell gene therapy.

Answer on p. 230

Exam tip

Remember that germline therapy will affect offspring and all future generations.

Revision activity

Write a list of all the key terms used in this chapter. Write the meaning of the key term next to it.

Exam practice

1 (a) Name the following:
   (i) an enzyme that synthesises new DNA [1]
   (ii) an enzyme that cuts DNA at a specific sequence [1]
   (iii) an enzyme that seals two pieces of DNA together [1]
   (iv) small, circular pieces of DNA in bacteria [1]
(b) Suggest a suitable vector for each of the following genetic engineering procedures:
   (i) transforming bacteria to express the gene for human insulin [1]
   (ii) engineering rice plants to express the enzyme to make vitamin A [1]
   (iii) treating cystic fibrosis by introducing working alleles to the lung cells of a human patient [1]
   (iv) cloning a large human gene inside bacteria [1]
(c) A–E represent steps in the genetic engineering process to make bacteria capable of producing human insulin. Place steps A–E in the correct order. [5]
   A use reverse transcriptase to make cDNA and add sticky ends
   B incubate plasmid with restriction enzyme
   C extract mRNA from human pancreatic cells
   D heat-shock bacteria in calcium chloride solution and add recombinant vector
   E incubate prepared gene with cut plasmid

2 (a) Haemoglobin is made up of the polypeptides α-globin and β-globin. Digestion of normal human DNA with the enzyme *HPaI* produces a fragment 7.6 kbp long containing the β-globin gene. Digestion of DNA from a person with sickle-cell anaemia produces a fragment 13 kbp long.
   (i) What type of enzyme is *HPaI*? [1]
   (ii) Describe how these fragments could be separated using gel electrophoresis. [4]
(b) (i) What is a gene probe? [2]
   (ii) Suggest how a gene probe can be used to screen members of a family with a genetic disorder such as muscular dystrophy. [5]

3 (a) During the polymerase chain reaction the temperature of the mixture is altered. List the temperatures involved and describe the process occurring at each temperature. [6]
(b) Explain the purpose of adding primers to the PCR mix. [3]

Answers and quick quiz 23 online

ONLINE

Summary

By the end of this chapter you should be able to:

- outline the principles in DNA sequencing and new sequencing techniques
- explain how gene sequencing allows for genome-wide comparisons between individuals and between species
- explain the role of DNA sequencing in prediction of amino acid sequences and synthetic biology
- outline DNA profiling
- explain what is meant by genetic engineering, including extraction of DNA, separation using electrophoresis, use of DNA probes, the polymerase chain reaction (PCR), use of vectors such as plasmids and how plasmids may be taken up by bacterial cells
- discuss the ethical concerns raised by the genetic manipulation of animals (including humans), plants and microorganisms
- explain the term gene therapy and the differences between somatic cell gene therapy and germ line cell gene therapy

24 Cloning and biotechnology

- Many plants can reproduce asexually to produce clones. Growers exploit this natural ability to reproduce by vegetative propagation to produce crops that are uniform and easy to harvest.
- We can now produce artificial clones of plants and animals. These can be used to increase numbers quickly and to reproduce combinations of characteristics that might be lost if sexual reproduction occurred.
- Biotechnology is the use of living organisms (or parts of living organisms) in industrial processes. Products such as food, drugs or other molecules can be manufactured.

Reproductive and non-reproductive cloning

A **clone** is an exact copy. In particular, it is a genetically identical copy. Cloning involves producing many genetically identical copies of individual organisms, cells or single genes.

A **clone** is an exact genetic copy.

Reproductive cloning is the term used for the process of artificially cloning whole organisms. This is used in biotechnology and in agriculture. Non reproductive cloning is the process of making many identical copies without the intent to produce a new organism. This could involve:

- making copies of a length of DNA by PCR for use in research
- growing stem cells for research or for treatment of disease
- growing tissue in tissue culture for research or for replacement of tissues or organs

To be genetically identical the cells and individuals must have been produced by cell division involving only mitosis.

Cloning plants

Production of natural clones

REVISED

Many plants reproduce asexually in nature. Plants reproduce asexually by vegetative propagation in the following ways:

- Producing runners or stolons, which are stems that grow along on the surface or just below the surface of the ground. Occasionally these horizontal stems take root and grow a new vertical stem. Examples include strawberries and spider plants.
- Root suckers or basal sprouts are similar, but rather than growing a horizontal stem the plant can produce new stems at intervals along the roots. These may form as a result of damage to the parent plant. Such suckers tend to grow in a circle around the old trunk — called a clonal patch. Elm trees produce root suckers.
- Bulbs and tubers, which overwinter underground. Examples include patches of daffodils produced from one original bulb and the growth of many potato tubers from one original tuber.

Tissue culture

REVISED

Many plants can reproduce by fragmentation — when a small part of the plant can regenerate a whole plant. Many gardeners increase the number of plants in their garden by taking cuttings — small branches or stems taken from a plant can take root when placed in compost. The success of these cuttings can be improved by dipping the cut ends in commercial rooting powder, which contains the plant growth regulator auxin.

Tissue culture is the cloning of cells from a small group of genetically identical cells to form a mass of similar cells.

Tissue culture is a modern development of this type of cloning. Much smaller pieces of plant tissue are placed on a special growth medium and caused to grow by mitotic cell division. They can be made to differentiate, forming new roots and leaves. This is also known as micropropagation. It involves a number of steps:

- The plant material is cut into small pieces called **explants**. These could be tiny pieces of leaf, stem, root or bud. The meristem is often used, because these cells are free from diseases.
- The explants are sterilised with bleach or alcohol to kill any bacteria and fungi. The following stages are then carried out under aseptic conditions.
- The sterilised explants are placed on a growth medium (agar jelly) containing nutrients such as glucose, amino acids and phosphates. They will also contain suitable high concentrations of plant growth substances such as auxin and cytokinin. This causes the cells in each explant to divide by mitosis to form a callus (a mass of undifferentiated cells, which are all **totipotent**).
- Each callus can then be subdivided, to increase the final number of plants that will be made.
- The callus is moved twice onto new growth media, each containing different ratios of plant growth substances. This makes the cells of the callus differentiate into tissues and organs to form a plant. (100 auxin : 1 cytokinin causes roots to form; 4 auxin : 1 cytokinin causes shoots to form).
- The tiny plantlets are then transferred to compost in a greenhouse where they can acclimatise to normal growing conditions.

An **explant** is a small piece of tissue that can be grown to produce a new plant.

A **totipotent** cell is capable of differentiating into any type of cell found in the organism.

Now test yourself

TESTED

1 Explain why it is essential to carry out this process under aseptic conditions.

Answer on p. 231

Typical mistake

Students forget that the callus can be further divided and that the procedure must be carried out aseptically.

Exam tip

The description of a technique such as this could be tested as part of How Science Works, and you could be asked to justify certain steps. Alternatively you could be asked to describe the process, with a mark awarded for sequencing the procedure correctly.

Advantages and disadvantages of cloning plants in agriculture

REVISED

Table 24.1 The advantages and disadvantages of cloning plants

Advantages	Disadvantages
Rapid way of producing new plants compared with growing plants from seed (some plants, e.g. orchids, are hard to grow from seed)	Tissue culture is an expensive process
New plants will be disease-free (no viruses)	Tissue culture can fail due to microbial contamination
Plants with specific characteristics can be selected and reproduced without forming hybrids that may have undesirable characteristics	All the cloned offspring are susceptible to the same pest or disease, so if one is infected they may all become infected; crops are grown in monocultures, which allows pests and disease spread easily
The new plants will be uniform — they all have same phenotype — making growing and harvesting easier	There is little or no genetic variation; the only source of variation in a clone is mutation, so the gene pool is reduced and evolution is unlikely to occur
Infertile plants such as triploids can be reproduced, such as commercially grown bananas	

Now test yourself

2 Explain why it is important to avoid sexual reproduction if you want to create exact genetic copies.

Answer on p. 231

Revision activity

Draw a mind map with 'cloning plants' at the centre. Include natural and artificial cloning and the techniques used to clone plants as well as advantages and disadvantages.

Cloning animals

Very few animal species naturally form clones but some invertebrate animals can reproduce asexually by budding or parthenogenesis. One notable exception is the production of identical twins in mammals (including humans). If a fertilised egg divides by mitosis and then the cells separate completely two new individuals will be produced. These two individuals will be genetically identical. This is known as embryo splitting.

Embryo splitting can also be used to artificially produce animal clones.

Embryo splitting

- A zygote is created (usually by in vitro fertilisation or IVF).
- It is allowed to divide by mitosis to form a ball of cells.
- These cells are separated before differentiation starts.
- Each cell is then allowed to grow and divide again.
- Each new group of cells can be placed into the uterus of a surrogate mother.
- This will produce many genetically identical offspring.

This technique can be used for reproductive cloning:

- to create many examples of elite farm animals that may have been produced by selective breeding or by genetic modification (genetic engineering)
- to produce animals for research, such as mice for testing new drugs

It can also be used for non-reproductive cloning:

- to produce genetically identical embryos for research into the action of genes that control development and differentiation
- to grow new tissues and organs as replacement 'spare parts' for people who are ill — tissues produced from a patient's own cells will be genetically identical and so will not be rejected by the immune system

Adult cell cloning

REVISED

This was first successful with Dolly the sheep in 1996. This technique is called somatic cell nuclear transfer (SCNT):

- An egg cell is harvested and its nucleus is sucked out using a micropipette.
- A normal body cell (somatic cell) from the adult to be cloned is harvested (for Dolly this cell was from the mammary gland of the original sheep).
- An electric shock triggers the enucleated egg cell and the somatic cell to fuse and start mitosis and development as though it had just been fertilised.
- The embryo is placed into the uterus of a surrogate mother.

Advantages and disadvantages of cloning animals

REVISED

The main problems with animal cloning are similar to the problems of plant cloning — a lack of genetic variability meaning that all the animals are equally susceptible to a certain disease, the gene pool is reduced and selection cannot take place. Cloned animals may also be less healthy and have shorter life spans.

Table 24.2 The advantages and disadvantages of cloning animals

Advantages	Disadvantages
High-value animals — those giving high yields or good quality meat — can be reproduced exactly	We are still not certain if the procedures involved cause other health issues
Rare or endangered animals can be reproduced for conservation purposes	Concern about the welfare of the animals
Genetically modified animals can be reproduced quickly — if the animals were modified to produce pharmaceutical chemicals this helps produce the drugs quickly	Lack of genetic diversity means an inability to adapt to changes in environment
	The exact genotype of organisms produced by embryo splitting depend on the sperm and egg used to create the zygote. The exact phenotype is not known until the cloned offspring are born
	The success rate of adult cell cloning is very poor and the method is expensive

Revision activity

Draw a mind map with 'cloning animals' at the centre. Include natural and artificial cloning and the techniques used to clone plants as well as advantages and disadvantages.

Biotechnology and microorganisms

Biotechnology is the use of microorganisms to carry out industrial processes such as:

- producing food:
  - Whole microorganisms are used in making cheese, yoghurt, bread, beer and single-cell proteins.
  - Enzymes are used in other processes, such as pectinase to release juice from fruit and glucose isomerase to make sweet products.
- producing medicinal drugs:
  - Whole organisms are used in manufacturing drugs such as penicillin and insulin.
  - Genetically modified mammals such as sheep and goats are also used to produce useful proteins, which are harvested in their milk.
- producing other products, such as methane or 'biogas' and citric acid, which is the food preservative E330

Microorgansisms are also used to treat waste products like waste water and sewage, and in bioremediation — cleaning up polluted areas.

Biotechnology is the use of organisms or parts of organisms (enzymes) in industrial processes.

Exam tip

Biotechnology is still a growth area and more applications are being found. This gives plenty of scope to test your understanding through using unfamiliar examples, so be ready to think about the example given and apply the principles.

Why microorganisms?

Microorganisms are not the only organisms used in biotechnology. However, microorganisms such as bacteria and fungi are often used because they:

- can be grown relatively cheaply
- grow rapidly even at low temperature
- have a short life cycle, so they reproduce quickly
- reproduce asexually, so that all individuals are genetically identical
- can produce proteins or other substances that are secreted into their external environment and can be isolated relatively easily
- can carry out quite complex processes, which would be difficult using chemicals
- can be genetically engineered
- can be easily selected for certain characteristics (strain selection)
- can be grown anywhere, making the process independent of climate
- can often be grown on waste products from other processes (e.g. molasses)

Revision activity

Write a list of all the topics here that link to other parts of the specification you have learnt about.

Using microorganisms to manufacture food

Yoghurt

REVISED

Milk is fermented by the action of *Lactobacillus bulgaricus* bacteria. This converts lactose to lactic acid. The low pH denatures proteins in the milk protein, causing it to coagulate, so it thickens. Fermentation also produces the flavours characteristic of yoghurt.

Cheese

REVISED

Milk is fermented by the action of bacteria (*Lactobacillus*) to produce lactic acid, which acidifies the milk. The enzyme rennin (chymosin) is added, which coagulates the milk protein (casein) in the presence of calcium ions. The resulting curd solid is separated from the liquid whey by cutting, stirring and heating. The curd is then pressed into moulds. The characteristics and flavour of the cheese are determined by its treatment while making and pressing the curd and further ripening.

Baking

REVISED

Flour is mixed with water, salt and yeast (a single-celled fungus, *Saccharomyces cerevisiae*). The dough is left in a warm place for a few hours while the yeast respires anaerobically to produce carbon dioxide bubbles, which make the dough to rise. The risen dough is baked.

Brewing

REVISED

S. cerevisiae is also used to make alcoholic beverages:

- Wine is made using grapes that naturally have yeasts on their skin. When the grapes are crushed, the yeast uses the sugars fructose and glucose in the grapes to produce carbon dioxide and alcohol.
- Beer is made using malted barley grains — these are beginning to germinate, converting stored starch to maltose, which is respired by the yeast. The anaerobic respiration produces carbon dioxide and alcohol.

Single-cell proteins

REVISED

Fungi such as *Fusarium venenatum* can be grown as food. These fungi manufacture a type of protein called mycoprotein. The best known example is Quorn™. Quorn™ is used as a healthy substitute for meat and to make vegetarian foods.

The advantages and disadvantages of using microorganisms to grow food are listed in Table 24.3.

Table 24.3 The advantages and disadvantages of using microorganisms to grow food

Advantages	Disadvantages
Production can be increased and decreased fairly quickly to match demand No animals are used, so vegetarians can eat the proteins There are no issues over how animals are kept The protein produced is healthier because it contains: • all the essential amino acids (it is a first-class protein) • less fat/saturated fat • no cholesterol The microorganisms can digest and grow on almost anything organic, so they can be grown on waste products from other processes or even on domestic waste The protein produced can be much cheaper than the animal-derived protein	Many people may not want to eat single-cell protein that has been grown by fungi; they may feel that it is unclean, or still contaminated People may not want to eat protein that has been grown on waste products The protein does not have the same taste or texture as traditional animal protein The protein contains less energy The protein needs to be isolated and purified before use The fermenters used maintain ideal conditions for all microorganism growth, so they may become infected with the wrong microorganisms or with pathogenic microorganisms, which will also grow quickly, making the production process less efficient or even making the product unusable

Other biotechnology processes

Penicillin

REVISED

Penicillin is produced through fermentation by the fungus *Penicillium chrysogenum*. A fermenter containing the fungus and all the nutrients required for its growth is run for 6–8 days. During this time the culture population increases in size until it is near its limits. Penicillin is a secondary metabolite — it is only produced once the population has reached a certain size. Once the fermentation is complete the culture is filtered to remove the cells. The antibiotic is precipitated and purified.

Insulin

REVISED

Insulin is used to treat type 1 diabetes. Synthetic human insulin was developed by genetically modifying the bacterium *Escherichia coli*. The resulting genetically modified bacterium is grown in fermenters to produce plentiful quantities of human insulin at relatively low cost.

Bioremediation

REVISED

Bioremediation is using microorganisms to clean up pollution. The microorganisms convert the toxic pollutants to less harmful substances. Crude oil, solvents and pesticides can be treated using bioremediation.

Bioremediation has certain advantages over other methods of clean up:

- It uses natural systems.
- Less labour and equipment are required.
- Treatment can be carried out on site.
- Few waste products are produced.
- There is less risk of harmful exposure to clean-up personnel.

Culturing microorganisms

In the laboratory, microorganisms are usually grown either in a liquid broth or on an agar plate. A broth is often used to allow the population to grow and samples are transferred to an agar plate in order to identify the microorganisms or count them.

Growing microorganisms on agar plates involves three main steps:

1. Sterilisation — making sure all the equipment is free from contamination. Most equipment can be sterilised in an autoclave, which heats water under pressure to 120°C.
2. Inoculation — transferring a sample of microorganisms to the surface of the agar. This can be done with a wire loop, as shown in Figure 24.1, or with a glass spreader.
3. Incubation — providing a suitable temperature for growth. Agar plates are placed upside-down in an incubator — this is to stop the agar drying out and to prevent drops from condensation disturbing the growth of new colonies.

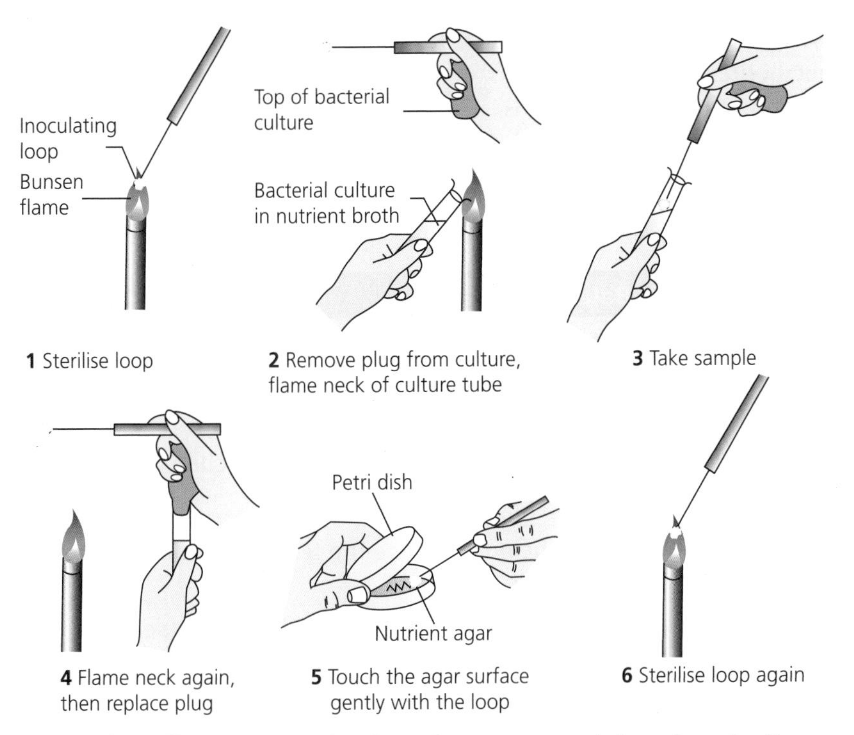
Inoculating loop
Bunsen flame
Top of bacterial culture
Bacterial culture in nutrient broth
1 Sterilise loop
2 Remove plug from culture, flame neck of culture tube
3 Take sample
Petri dish
Nutrient agar
4 Flame neck again, then replace plug
5 Touch the agar surface gently with the loop
6 Sterilise loop again

Figure 24.1 The steps used to inoculate an agar plate using sterile technique

Aseptic techniques

REVISED

Aseptic techniques are precautions taken to avoid unwanted contamination. Standard procedure is as follows:

1. Wash your hands thoroughly.
2. Disinfect the working area with a suitable disinfectant or alcohol.
3. Light a Bunsen burner and keep it on — this is used to sterilise the implements but also heats the air, causing it to rise so that air-borne microorganisms do not settle.
4. As a bottle is opened its neck is passed through the flame to prevent bacteria in the air entering the bottle. This should also be done when the top is returned to the bottle.
5. Lift the lid of the Petri dish only enough to allow introduction of the desired microorganism.
6. Close the Petri dish and use tape to keep the dish closed.
7. Any glassware or metal equipment should also be passed through the flame before and after use.
8. Wash your hands again.

Using fermenters

For industrial processes the growth of microorganisms must be scaled up. This is achieved by using a fermenter to grow the culture. A fermenter is a large vat or vessel in which the correct conditions for microorganism growth can be maintained.

Continuous and batch culture

REVISED

Microorganisms can be grown in two ways:

- Continuous culture is where a culture is set up and nutrients are added and products are removed from the culture at intervals. The culture

is maintained at the exponential phase of the growth curve so that it continues to grow and produce its metabolites quickly.
- Batch culture is where a starter population of microorganisms is supplied with a fixed amount of nutrients and allowed to grow. At the end of the time period the products are extracted.

Table 24.4 Comparing continuous and batch culture

Continuous culture	Batch culture
Nutrients added continuously	Nutrients added at start only
Maintaining the correct conditions can be difficult and expensive	Can be left to continue for a set time period
High growth rate as nutrient levels maintained	Slower growth rate as nutrients decline
Products collected continuously	Products collected at end
Fermenter is in use continuously, which is more efficient	Less efficient because the fermenter is not in use constantly
The microorganisms are metabolising normally — this is useful for producing primary metabolites	Useful for secondary metabolites as these are generated when population growth rate declines
Examples include production of insulin and single-cell protein	Examples include wine, beer and yoghurt
In the event of contamination production is stopped and losses can be great	In the event of contamination only one batch is lost

Revision activity

Draw a fermenter and annotate the diagram with notes explaining how and why the conditions are maintained.

Primary and secondary metabolites

REVISED

Primary **metabolites** are produced in the course of normal metabolism. All microorganisms produce primary metabolites. Examples include proteins, enzymes and alcohol.

Secondary metabolites are produced after the main population growth has occurred, nutrients are in short supply and the population is not growing rapidly. Examples include antibiotics.

Metabolites are substances produced by living organisms.

Now test yourself

TESTED

3 Explain why batch fermentation is best for harvesting secondary metabolites.

Answer on p. 231

Typical mistake

Try not to confuse similar terms such as primary metabolites and secondary metabolites, continuous culture and batch culture.

Maximising the yield

REVISED

A fermenter can be used to optimise the growing conditions so that the maximum yield is produced. The following conditions are maintained or even altered during the fermentation process:
- Temperature — it needs to be warm enough to maintain the metabolic rate but not hot enough to denature the enzymes. Once the fermentation is in process the microorganisms produce their own heat and it may be necessary to cool the fermentation vessel with cold water.

- Substrate or nutrients — the correct nutrients must be added, including sources of carbon and nitrogen.
- Oxygen — oxygen is required when processes involving aerobic respiration are being used.
- pH — this must be maintained to ensure enzyme action continues.
- Agitation — if the culture is not stirred the microorganisms may settle to the bottom of the fermenter. The culture can be agitated by a rotating paddle or by bubbling oxygen in at the base.

Exam tip

Remember that enzymes are involved, so all conditions need to be suitable for enzymes. There are obvious links here for testing your knowledge synoptically.

Now test yourself TESTED

4 Suggest how continuous culture could be used to produce secondary metabolites such as antibiotics.

Answer on p. 231

The importance of asepsis

REVISED

Nutrients, oxygen and an ideal temperature are provided in a fermenter, so the microorganisms are growing in ideal conditions. It is important to exclude other microorganisms because they would also grow well. Unwanted microorganisms could:

- compete for nutrients, reducing growth rate and yield
- kill the culture microorganism
- use the required product for its own metabolism
- spoil the product by contaminating it with toxic by-products

Asepsis involves keeping the culture free from unwanted microorganisms.

Asepsis means avoiding contamination by other microorganisms. It is achieved by:

- washing, disinfecting and steam cleaning all equipment — including the fermenter
- using a fermenter made from polished stainless steel, which prevents microbes sticking to surfaces
- sterilising all nutrients before they are added to the fermenter — this is usually achieved by steam or heat treatment
- making sure that any air or oxygen bubbled into the fermenter is free from microorganisms — this is achieved using very fine filters

Now test yourself

5 Explain why a fermenter must be thoroughly cleaned and disinfected between batches.

Answer on p. 231

TESTED

The standard growth curve

REVISED

The standard growth curve for a microorganism population is called a sigmoid curve (Figure 24.2). In a closed culture it has four phases:

- Lag phase — reproduction and growth are very slow as cells acclimatise, absorb nutrients, produce enzymes and store energy. This may involve activating genes to produce specific enzymes.
- Exponential or log phase — reproduction is rapid, cells may divide every 20–30 minutes and the population could double with every generation. There are no limiting conditions and few cells die.
- Stationary phase — the population remains constant as death rate equals reproduction rate.
- Decline phase — the death rate exceeds the reproduction rate often because there is some limiting condition such as:
  - high temperature
  - limited nutrients or oxygen
  - a build up of waste products, such as carbon dioxide or ethanol

Typical mistake

Some students confuse a normal growth curve showing the population size with a rate of growth curve.

Revision activity

Sketch a growth curve and annotate it with notes that explain the shape.

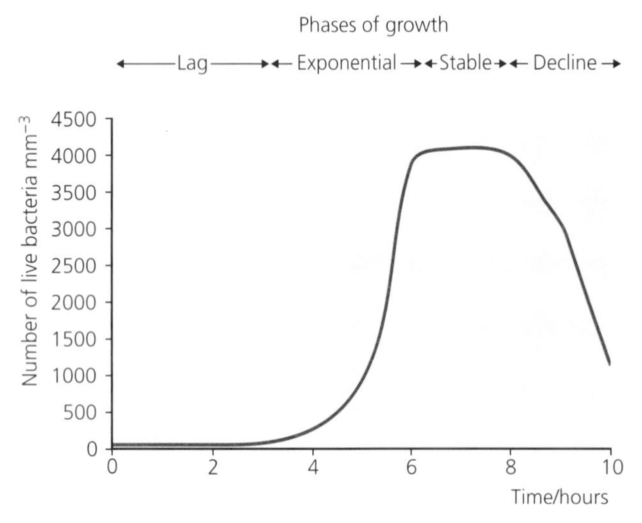

Figure 24.2 The standard growth curve of a population in a closed culture

Now test yourself

6 Explain why the growth curve comes to a plateau.

Answer on p. 231

TESTED

Immobilised enzymes

Enzymes alone can be used in many industrial processes. Enzymes can be fixed in place so that they do not mix freely with the substrate and are not lost in the process. There are several ways to immobilise enzymes:

- adsorption onto a surface such as glass, carbon or clay; the enzymes are held in place on the surface by hydrophobic or ionic interactions
- covalent bonding of enzymes to a surface
- entrapment in gel beads or between fibres of cellulose
- separation from the substrate by a partially permeable membrane

Immobilised enzymes have many advantages in large-scale production:

- Enzymes are very good catalysts.
- Enzymes are specific to the process — so there are few by-products that would need to be removed from the final product.
- Reactions can be carried out at fairly low temperatures — around 30–70°C — which saves heating costs.
- The enzymes do not contaminate the end product.
- The enzymes are not lost and can be reused straight away.
- The enzymes are often protected by the immobilising matrix, so they are less likely to be damaged by high temperature or extremes of pH.

Table 24.5 lists examples of immobilised enzymes used in industry

Immobilised enzymes are enzymes that are fixed to a surface so that they do not freely mix with the substrate.

Revision activity

Draw a mind map with 'immobilised enzymes' at the centre. Include all the information about how enzymes can be immobilised, enzyme activity and the effects of surrounding conditions.

Revision activity

Write a list of all the key terms used in this chapter. Write the meaning of the key term next to each one.

Table 24.5 Immobilised enzymes used in industry

Enzyme used	Industrial application
Glucose isomerase	• Converts glucose to fructose, producing high-fructose corn syrup (HFCS) • HFCS is sweeter than sucrose and is used in 'diet foods'
Penicillin acylase	• Manufacture of synthetic penicillins such as amoxicillin and ampicillin • Some resistant microorganisms are not resistant to synthetic penicillins
Lactase	• Converts lactose to glucose and galactose in lactose-free milk • People who are lactose intolerant can use lactose-free milk as a good source of calcium, to reduce the risks of weak bones or osteoporosis
Aminoacylase	• Used to produce pure samples of L-amino acids, which are used as the building blocks for synthesis of a number of pharmaceutical molecules
Glucoamylase	• Converts short-chain carbohydrates (dextrins) to glucose • Used to convert raw starch pulp to glucose, which can be fermented to produce gasohol or other products

Exam practice

1 (a) (i) Explain what is meant by non-reproductive cloning. [2]
(ii) Describe two uses for non-reproductive clones. [2]
(b) (i) Explain what is meant by reproductive cloning. [1]
(ii) Describe *three* advantages of plant tissue culture. [3]
(iii) Dolly the sheep was cloned by adult somatic cell nuclear transfer. Outline the technique used to clone a mammal from an adult cell. [4]

2 (a) Potatoes are rarely grown from seed. The usual method is to save some of the harvested potatoes and to re-plant them in the spring.
(i) State two advantages of growing potatoes in this way. [2]
(ii) In Ireland potato blight caused by the fungus-like organism *Phytophthora infestans* caused widespread destruction of the potato crops between 1845 and 1850. Suggest what made the potato crop so susceptible to the fungus. [3]
(b) Modern potato stocks can be certified free from disease.
(i) Describe how a blight-resistant potato plant could be produced. [3]
(ii) Describe how large numbers of resistant plants could be produced quickly. [4]

3 (a) High-fructose corn syrups are produced using an immobilised enzyme called glucose isomerase.
(i) Suggest how an enzyme can be immobilised for a student investigation. [2]
(ii) Describe two advantages of immobilising enzymes in an industrial process. [2]
(iii) Fructose and glucose are both monosaccharides with the formula $C_6H_{12}O_6$. Suggest what is the action of glucose isomerase. [2]
(b) (i) What is a fermenter? [2]
(ii) State two conditions inside a fermenter that are controlled and explain why it is necessary to control each. [4]
(iii) Describe how a fermenter is kept aseptic. [4]
(c) Compare and contrast the production of primary and secondary metabolites in a fermenter. [7]

Answers and quick quiz 24 online

ONLINE

Summary

By the end of this chapter you should be able to:

- describe the production of natural clones in plants using the example of vegetative propagation
- describe the production of artificial clones of plants and tissue culture used in micropropagation
- discuss the advantages and disadvantages of plant cloning in agriculture
- discuss arguments for and against artificial cloning in plants
- describe how natural clones of animals are produced
- describe how artificial clones of animals can be produced
- discuss the advantages and disadvantages of cloning animals
- discuss arguments for and against artificial cloning in animals
- describe the use of microorganisms in biotechnological processes
- explain why microorganisms are often used in biotechnological processes
- discuss the the advantages and disadvantages of using microorganisms to make food
- describe how microorganisms can be cultured using aseptic techniques
- describe the use of fermenters in continuous culture and batch culture
- describe the differences between primary and secondary metabolites
- explain the importance of manipulating the growing conditions in a fermentation vessel in order to maximise the yield of product required
- explain the importance of asepsis in the manipulation of microorganisms
- describe, with the aid of diagrams, and explain the standard growth curve of a microorganism in a closed culture
- describe how enzymes can be immobilised
- explain why immobilised enzymes are used in large-scale production
- describe the uses of immobilised enzymes in industry

25 Ecosystems

Organisms have complex interactions with other organisms and also with their environment. As a result, ecosystems are dynamic and change usually progresses towards a climax community. Biomass is transferred from one trophic level to another and the efficiency of this transfer limits the number of organisms that can live in a particular ecosystem.

What is an ecosystem?

An **ecosystem** comprises all the living things in one area, their interactions with each other and their interactions with their environment, including factors in the soil. Ecosystems vary in size from a rock pool on the seashore to a huge forest. The edges of ecosystems are not clearly defined — one might merge into another.

An **ecosystem** describes all the interactions between the living and non-living components in a defined area.

Factors that affect ecosystems can be biotic or abiotic.

Biotic factors

REVISED

These are factors that involve other living things, such as predation, food supply, disease and competition.

Abiotic factors

REVISED

These are non-living factors, such as pH, temperature, moisture/rainfall/relative humidity, wind, soil type and concentration of pollutants.

Typical mistake

Many students forget about the abiotic factors when they are describing an ecosystem.

Organisation of ecosystems

The ecosystem is divided into trophic (feeding) levels:

- **Producers** are the lowest trophic level. They are autotrophs that can convert energy from the environment into chemical energy in the form of complex organic molecules. These molecules are then used for growth or as substrates in respiration. Producers include:
  - plants, which are phototrophs — they use the Sun's energy to convert small inorganic molecules into carbohydrates
  - certain bacteria that can use chemical energy and heat to convert small inorganic molecules into complex organic molecules — these are called chemotrophs
- **Consumers** occupy the higher tropic levels. They digest the complex organic molecules made by autotrophs and then use the products for growth or as substrates in respiration.
- Primary consumers feed on plants.
- Secondary consumers feed on primary consumers.
- Tertiary consumers feed on secondary consumers.
- Decomposers are organisms that feed on waste or dead organic matter. They gain their energy by digesting and respiring the complex molecules in the organic matter. They cause decay and lead to food going off.

Producers convert inorganic matter to complex organic molecules.

Consumers are organisms that feed on other organisms.

Revision activity

Draw a food web of a simple ecosystem — remember to keep organisms in trophic levels. At the side of each trophic level explain what role the organisms in the trophic level play in the ecosystem.

Ecosystems as dynamic systems

Ecosystems are not static — they are always changing. They are complex, with so many interlaced interactions that one change, however small, will cause other changes. Therefore ecosystems are called dynamic. Any factor that affects the way an organism lives or makes food easier to access or more difficult to find will cause some change in the ecosystem. Such changes alter the flow of biomass through the ecosystem.

Biomass transfer through an ecosystem

REVISED

Biomass is the mass of tissue that has come from living things (it may be still alive or it may have recently died). All living things need energy and building blocks for growth. Energy comes from the Sun and passes through the ecosystem. Materials that make up the building blocks for growth and biomass are constantly recycled as they pass from one trophic level to another before decomposition returns them to the start.

Biomass created through photosynthesis in plants is transferred from one trophic level to the next by feeding. The chain of biomass transfer is called a food chain and a number of food chains may interact to produce a food web (Figure 25.1).

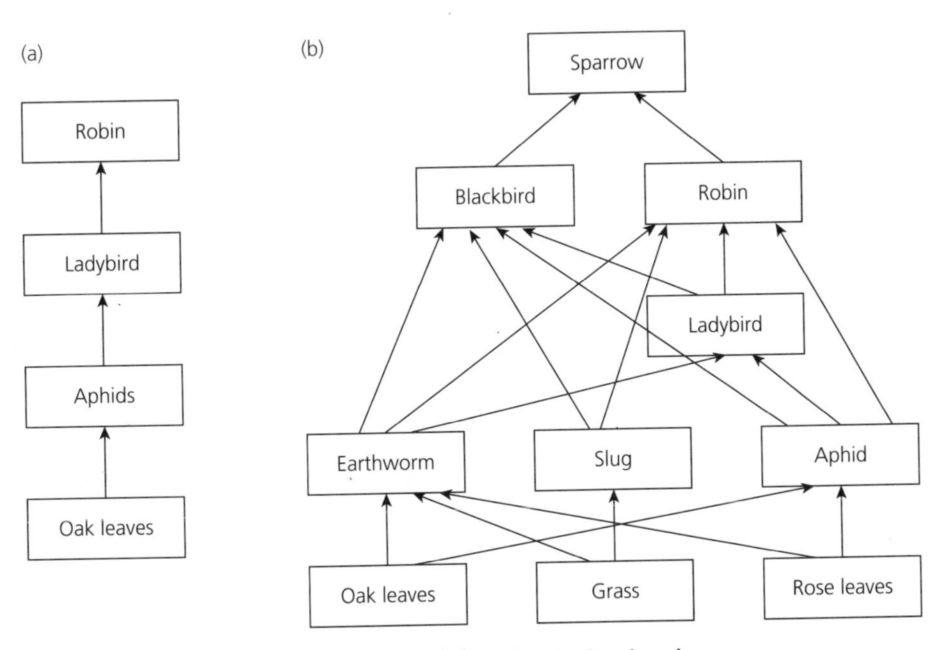

Figure 25.1 (a) A simple food chain (b) A simple food web

Loss of biomass

REVISED

At each tropic level some of the biomass is used for cellular respiration — the energy in the biomass is released when organic molecules in the biomass are converted to inorganic molecules such as carbon dioxide and water.

Biomass is also lost as it is converted to substances that may not be available to the next trophic level. Waste materials, dead organisms and molecules that cannot be digested will be available to decomposers (bacteria and fungi), which convert these materials back to inorganic molecules and return them to the soil or air. This is recycling.

As a result of biomass lost between trophic levels the higher levels contain less biomass. This is shown as a pyramid of biomass (Figure 25.2).

Top carnivore
Carnivores
Herbivores
Plants

Figure 25.2 A pyramid of biomass

Measuring biomass and the efficiency of biomass transfer

REVISED

The biomass in a trophic level can be measured by estimating the number of individuals in the trophic level and multiplying by the mean mass of individuals at that level. It is best to use the dry mass rather than the wet mass as the proportion of water in different organisms can vary.

The efficiency of biomass transfer between trophic levels is calculated by the equation:

$$\text{efficiency} = \frac{\text{biomass at higher level}}{\text{biomass at lower level}} \times 100$$

Typical mistake

Many students find even simple calculations very difficult. In this case some are unsure what to put as the lower figure in the equation.

Variation in biomass transfer

REVISED

The transfer of biomass between trophic levels is variable. This is because of the losses at each level and between levels. Losses include:

- energy released from biomass during respiration to power life processes such as movement and active transport
- tissue that cannot be eaten or digested easily, such as cellulose and bone
- energy held in the substances excreted such as carbon dioxide and urea
- organisms that grow but are not eaten by organisms in the next level

The losses between tropic levels are variable and some ecosystems are more efficient at passing biomass between trophic levels than others. There are many reasons for this:

- Ectotherms use less energy in keeping warm than endotherms, so more of the energy consumed can be used in growth of biomass and is available to the next tropic level.
- In some ecosystems a higher proportion of the available food is eaten.
- Some tissues are easier to digest than others.

Exam tip

Remember that net productivity is the gross productivity minus the losses due to respiration.

Human influence on biomass transfer

Humans have learned to manipulate the flow of biomass through ecosystems to improve the growth of food. Principally this involves diverting as much energy into growth of food as possible and creating artificial ecosystems. This may involve:

- improving the conditions for growth, using:
  - greenhouses where crops can be grown in artificial conditions to enhance their growth — extra light, warmer temperatures and extra carbon dioxide all help
  - irrigation
  - fertilisers
  - crop rotation
- improving the efficiency of energy conversion to food, by:
  - using faster-growing crops
  - selective breeding or genetic engineering to introduce fungus resistance or disease resistance in plants or cause faster growth/greater production in animals
  - harvesting animals as soon as they are full grown
  - keeping animals in small enclosures or even indoors to reduce loss from respiration used to move around and keep warm
- reducing competition, by:
  - controlling pests
  - using selective weedkillers
  - using antibiotics to treat illness

Recycling within ecosystems

Decomposers are microorganisms that cause decay. They release enzymes into their surroundings and absorb the small organic molecules that are produced by digestion. Decomposers are essential for breaking down all the large organic molecules that have been produced by autotrophs and heterotrophs. This includes:

- uneaten plants and animals that die
- parts of organisms that are shed — this includes leaves and exoskeletons
- the remains of plants and animals that have been partly eaten
- undigested material egested as faeces
- compounds that have been excreted

The decomposition of these large complex molecules is essential to return the simple inorganic components to the soil so that they can be used again by plants.

Decomposers are microorganisms that cause decay and recycle minerals.

Now test yourself

1 Suggest what might happen if the rate of decomposition was too slow.

Answer on p. 231

TESTED

Recycling of nitrogen

REVISED

The element nitrogen is essential for production of amino acids and proteins as well as for DNA. The recycling of nitrogen is shown in Figure 25.3.

Bacteria play an essential role in recycling nitrogen:

- *Nitrosomonas* and *Nitrobacter* are known as nitrifying bacteria. They are involved in converting complex nitrogen-containing compounds to simple inorganic compounds that can be absorbed by plants.
- *Azotobacter* is a free-living bacterium that fixes nitrogen.

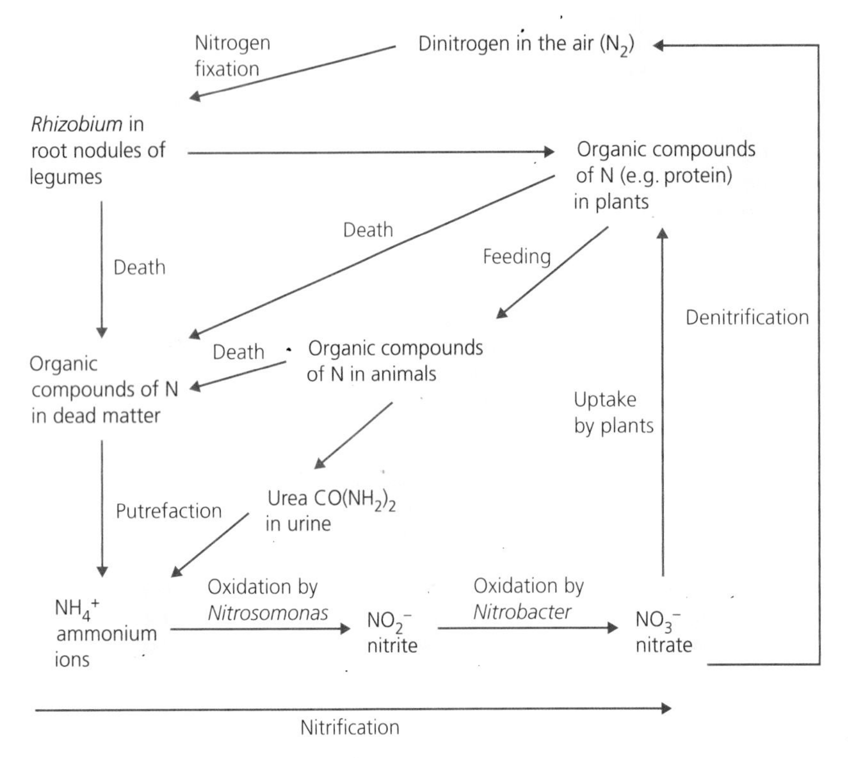

Figure 25.3 The nitrogen cycle

- *Rhizobium* is also a nitrogen-fixing bacterium. It lives free in the soil but also forms an association with plants in the legume family. It lives in special nodules in the roots of legumes and supplies the plant with ammonium ions.

Exam tip

Every book contains a different version of the nitrogen cycle. It is better to understand the cycle and be able to draw it by thinking about the processes involved than to learn any one diagram.

Now test yourself

TESTED

2 Explain why farmers in many parts of the world plant crops of beans or grass and clover every few years.

Answer on p. 231

Revision activity

Sketch a diagram of the nitrogen cycle, writing notes beside each arrow to explain the stage in the process it represents.

Recycling of carbon

REVISED

Carbon is another element that is recycled within ecosystems. The carbon cycle is shown in Figure 25.4.

The carbon cycle is also driven by the action of organisms. Fixing carbon from the air is achieved by green plants as they photosynthesise. Much of this carbon becomes part of the plant biomass, which is passed to animals through feeding.

Both plants and animals respire, returning carbon to the air as carbon dioxide. Death and the release of other materials such as excreta release carbon-containing compounds to the decomposers. This also returns the carbon to the air as carbon dioxide.

Some organic remains do not decompose and become fossilised. The carbon in fossils remains fixed until some process allows its release. This release is often associated with human activity, such as burning fossil fuels and using calcium carbonate in industrial processes.

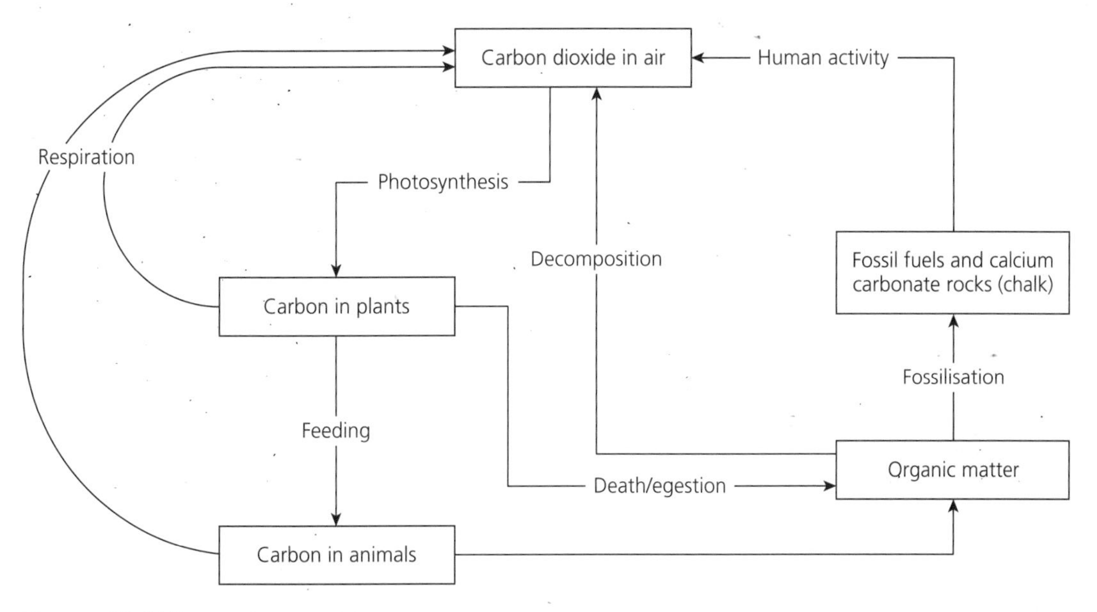

Figure 25.4 The carbon cycle

Succession

Succession is a directional change in the composition of a community.

Succession is the term used to describe the way that a community changes over time. Each new species living in a habitat will modify that habitat by changing the conditions slightly. As a result, new species may be able to migrate in and live successfully. As the conditions change species that have lived there successfully may be outcompeted by newer arrivals. Therefore the community living in that habitat changes over time. As a general rule the community becomes more complex as succession continues. Also, the organisms able to survive tend to get larger.

Primary succession is the sequence of organisms that live in an area starting from a newly created piece of land that has not been occupied previously. Such succession starts with pioneer species that are hardy and able to survive in harsh conditions. These colonisers are usually small and few in number. The succession continues as each new arrival modifies the environment and the community becomes more complex, with larger numbers of larger species. It continues until the climax community is reached — this is the final community, full of well-adapted competitors, and can remain unchanged for many years.

One example of primary succession is seen on sand dunes:

bare sand → sea rocket → marram grass → bird's-foot trefoil → clover → grasses (sand fescue) → small shrubs (bramble) → small trees (elder) → larger trees (silver birch)

Deflected succession

REVISED

If the conditions in an ecosystem are not suitable for the growth of the species found in the normal climax community then succession may be halted or altered. When human activity (either directly or indirectly) halts the normal succession it is known as deflected succession. This can occur as a result of the following activities: cutting down trees, mowing, grazing, burning, application of fertiliser and application of herbicide. This treatment produces a sub-climax community or plagioclimax.

Now test yourself

TESTED

3 Explain why the most diverse community is found just before the climax community is reached.

Answer on p. 231

Measuring distribution and abundance of organisms

The distribution and abundance of organisms can be measured in a number of ways:

- A line transect is used where the conditions change and the distribution of organisms also changes. It comprises a long tape measure or rope laid out on the ground. Observations are made along the transect and every species touching the transect is recorded, along with an estimate of its abundance. In a long transect observations might be made at suitable intervals.
- A belt transect is similar to a line transect but is a band rather than a single line. This means that each unit length of the belt transect provides an area in which abundance or percentage cover can be measured accurately rather than estimated.
- The use of quadrats is described on page 100

Exam tip

You do not need to remember a large number of examples of succession. Learn one example and remember that succession starts with pioneer species and the community becomes more complex, with larger species.

Revision activity

Look back at your work from AS and the ways in which biodiversity was measured.

Exam tip

This is an obvious topic for synoptic testing of material covered at AS.

Revision activity

Write a list of all the key terms used in this chapter. Write the meaning of the key term next to each one.

Exam practice

1 (a) (i) What is meant by the term 'producer'? [1]
(ii) Explain why a producer is only able to use about 1% of the Sun's energy. [4]
(b) The transfer of biomass between producers and primary consumers can be as low as 5%. Describe and explain how farmers improve the efficiency of biomass transfer between these trophic levels. [7]

2 (a) Explain the role of decomposers in an ecosystem. [3]
(b) Explain the reasoning behind the following statements:
(i) Sowing clover every fourth year improves productivity. [2]
(ii) Ploughing in the roots and stems of crops after harvest improves the soil. [2]
(iii) Maintaining hedgerows and wild areas on farms reduces damage to crops by pests. [2]
(iv) More people can be fed per hectare of land if they eat vegetarian food rather than meat. [2]
(v) Planting native tree species improves biodiversity more than planting imported species. [2]

Answers and quick quiz 25 online

ONLINE

Summary

By the end of this chapter you should be able to:

- describe ecosystems as dynamic and influenced by both biotic and abiotic factors
- describe how biomass is transferred though ecosystems, outline how biomass transfers between trophic levels can be measured and discuss the efficiency of biomass transfer between trophic levels
- explain how human activity can manipulate the flow of biomass through ecosystems
- describe the role of decomposers in the decomposition of organic material and describe how microorganisms recycle nitrogen and carbon within ecosystems
- describe one example of primary succession resulting in a climax community
- describe how the distribution and abundance of organisms can be measured, using line transects, belt transects, quadrats and point quadrats

26 Populations and sustainability

Factors that affect the size of a population

A population of organisms in a new environment will grow exponentially as long as there are no limitations to growth. However, eventually some **limiting factor** will determine its final size (Figure 26.1).

A **limiting factor** is some component of the ecosystem that limits the growth and final size of the population.

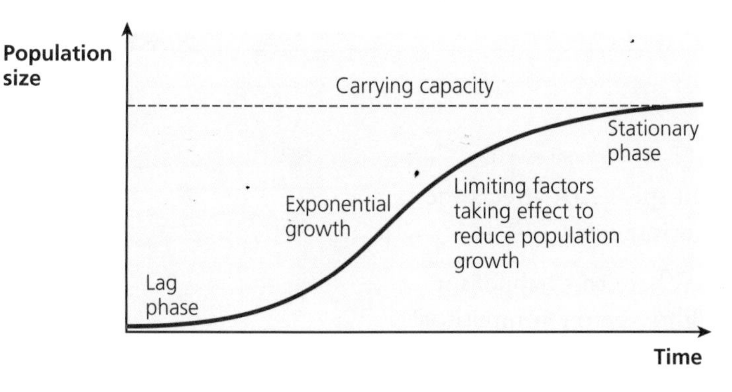

Figure 26.1 A typical population growth curve

The eventual maximum size of a population that can be sustained by its environment is known as the carrying capacity. The carrying capacity may be reached by a smooth curve or the population might overshoot and fall back to the final size. Limiting factors include:

- availability of resources, such as food, water, light (for plants), oxygen, space or suitable nest sites
- number of predators
- disease — in order to survive, the pathogen causing a disease must be transmitted from host to host. At a high density the pathogen can easily be transmitted and will cause disease, reducing the population size.
- behaviour — as population density increases, greater aggression or migration can limit population size (e.g. lemmings).
- natural disasters
- climate and weather

Revision activity

Draw a mind map with population size at the centre. Describe and explain the effects of all the factors that affect the size of a population.

Interactions between populations

Predator–prey interaction

REVISED

Predators feed on prey and reduce the prey population size. As they feed, the predator population will grow but if they over-use the prey population (eating too many) then the predator population will fall again as the prey population falls. This can cause a cyclical effect — as an increasing population of predators eats more prey the prey population drops followed by the predator population dropping. Once the predator population has dropped the prey population can grow again (e.g. *Typhlodromus* (predator) and *Phytonemus* (prey)(see figure 26.2)).

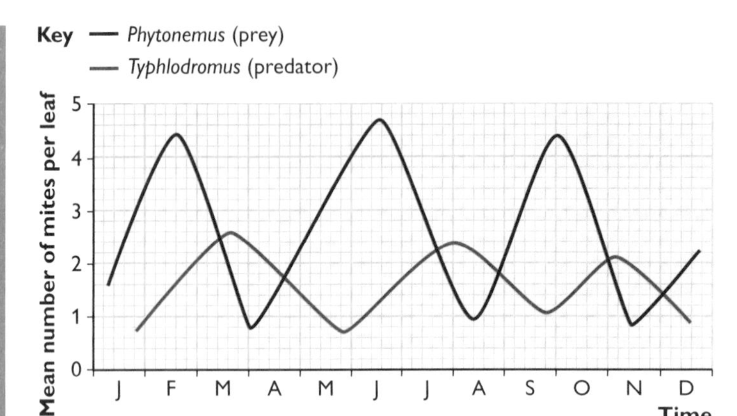

Figure 26.2 The cycles produced by predator–prey relationships

Interspecific and intraspecific competition

REVISED

Competition occurs when resources are limited.

Interspecific competition

This is competition between individuals of different species. It affects the sizes of both populations and the distributions of both species.

Two species cannot live in complete competition. Where this happens it is likely that one species will adapt to become a slightly better competitor and will out-compete the other. This may lead to:

- extinction of one species
- greater specialisation of each species to avoid competition — this could involve concentrating on one type of food or using a specific type of nest site
- a change in the distribution of the species — so they avoid competition by living in different areas. This may be because conditions in the two areas are different and give a slight advantage to one of the species. For example, in most of England the grey squirrel has out-competed the red squirrel. However, the red squirrel still survives and out-competes the grey in coniferous woodland.

Intraspecific competition

This is competition between individuals of the same species. As a population grows, resources become increasingly limiting and competition increases, so that only the best adapted — the best competitors — survive. This reduces the growth and limits the eventual size of the population.

Exam tip

Don't forget that competition only occurs once resources are limited and the effect of competition on evolution is important — it allows natural selection to occur.

Now test yourself

TESTED

1 Suggest why scramble competition rather than a dominance hierarchy is more likely to lead to natural selection.

Answer on p. 231

Conservation and preservation

Conservation and preservation both seek to overcome threats to biodiversity, many of which result from human activity. The threats to biodiversity include:

- over-exploitation of wild populations for food (e.g. overfishing)
- habitat destruction for agriculture or development
- habitat disruption and fragmentation as a result of development
- introduction of non-native species that out compete the native species

Preservation is maintaining or protecting an area in its current state. This is best applied to areas that have not yet been affected by human activity. If the area has already been affected it may not have the suitable environmental conditions to enable survival of the ecosystem.

Conservation is an active process in which positive action is taken to maintain and enhance biodiversity. This may include:

- setting up reserves such as national parks, national nature reserves and sites of special scientific interest
- legal protection
- education of the local population
- active breeding programmes
- importing individuals from captive breeding programmes or from other more stable ecosystems to increase population size
- removal of excess predators
- prevention of poaching
- monitoring of health and vaccination to reduce disease
- provision of food to reduce competition
- active management to provide suitable diverse habitats
- reclamation of land in order to return it to the original habitat and ease competition for space

Conservation is an active process, seeking to increase biodiversity.

Economic, social and ethical reasons for conservation

REVISED

Conservation programmes concentrate on maintaining biodiversity. There are several reasons for conservation:

- Every species has its own value — and a right to survive. Since many species are endangered as a result of human activity we have an ethical responsibility to conserve them.
- Many species have an economic value when harvested as a food source or fuel supply.
- Many species have a future potential as a source of genetic diversity for breeding new characteristics into our crops or as a source of medicinal drugs.
- Provision of natural predators controls the pests that damage crops.
- Pollination of crops requires a healthy insect population.
- Provision of so-called environmental services, such as:
  - maintenance of water quality
  - maintaining the water cycle
  - maintenance of soil quality
  - recycling minerals
  - decomposition of waste
- Ecotourism brings wealth to some areas.

Typical mistake

This is another example of a topic where some students write inappropriate, lengthy arguments about the ethics and morals of human activities. At A-level you need to keep to the facts.

Revision activity

Draw a mind map with 'conservation' at the centre. Include all the reasons for conservation, grouping the reasons according to whether they are economic, social or ethical.

Now test yourself

TESTED

2 Explain why preservation is not enough to maintain species diversity.

Answer on p. 231

Sustainable management of ecosystems

An ecosystem managed in a **sustainable** way can continue to produce crops indefinitely. Sustainable management of an ecosystem can achieve a wide range of goals that may, at first, appear to conflict. For example, temperate woodland can be managed to produce a sustainable harvest while still allowing amenity use and maintaining biodiversity.

Sustainable means that harvesting can be carried out indefinitely without damaging the ecosystem.

Sustainable management of woodland

REVISED

This involves a number of simple approaches:

- Never clear fell; use selective thinning or selective felling so that the ground is never fully exposed over a large area.
- Replace felled trees using fast-growing native species.
- Protect young trees from browsing by rabbits and deer.
- Harvest timber from a different part of the woodland each year and rotate the areas used to avoid clearing one huge area. This practice also means that a wide range of habitats is maintained as each year new habitats are created by harvesting.
- Coppicing — cutting the plants close to ground level, leaving the roots so that the soil is not disturbed. This also encourages more growth and produces timber that has a variety of traditional uses or can be used as fuel. The cut branches can also be stacked and allowed to decay slowly, which provides a range of habitats for many small birds, mammals and invertebrates.
- Pollarding — cutting all the branches off taller trees at 3–6 m above the ground. This has the same advantages as coppicing.
- Use of standards — these are tall trees left to grow between the coppiced plants, which protect soil. These trees can also be harvested once they are large enough.
- Include rides or fire breaks to protect the crop from fires. These also double as walking trails or cycle tracks.
- Incorporate other land uses in the woodland such as:
  - paths for walkers or off-road cyclists
  - leisure activities such as camping, orienteering or paint-balling
  - conservation — leave areas of scrub/piles of dead wood for habitats

Typical mistake

Some students forget to mention the rotational aspect of harvesting — this is perhaps the most important factor.

Now test yourself

TESTED

3 Explain why rotational coppicing is a good management technique.

Answer on p. 231

Sustainable fishing

REVISED

Sustainable fishing means that fishing must take place at a level that allows it to continue indefinitely. Overfishing lowers the fish population to the extent that harvests are reduced and the fish population could be lost altogether.

Fishing is managed to maintain the structure, productivity, function and diversity of the ecosystem so that there is no permanent damage to the local habitat and dependent species. Optimal sustainable fishing aims to maintain the fish population at its carrying capacity while harvesting any fish in excess of the carrying capacity for that environment. Fish farms, or aquaculture, can produce fish efficiently without affecting wild populations in the oceans.

The effects of human activities

Human activities have an effect on the local ecosystem. However, ecosystems can be managed to balance the conflict between conservation/preservation and human needs. This is possible even in relatively sensitive ecosystems.

The Masai Mara region (Kenya)

REVISED

The national reserve has successfully combined the needs of the local people to run farms for food production with larger land owners who have modified their land use to encourage conservation and generate income from ecotourism.

The Terai region (Nepal)

REVISED

Marshy grassland, savannah and forest is home to endangered species such as the Bengal tiger. The local people are heavily reliant upon the forest, which is under pressure from increased agriculture and grazing. Community forestry initiatives give local people rights to exploit the forest in return for adopting responsibility to look after it. The local people benefit through marketing products made from the forest.

Peat bogs

REVISED

Peat accumulated over thousands of years contains lots of historical archaeology. Soil conditions are ideal for *Sphagnum* moss forming some of the UK's most scarce wetland habitats important as feeding and stopping-off points for migrating birds. The UK Biodiversity Action Plan (UKBAP) aims to conserve and enhance biodiversity through local-level schemes involving the RSPB and English Nature by restoring certain peat bogs and ending the commercial use of peat in the UK.

The Galápagos Islands

REVISED

Since Charles Darwin's discoveries made the islands famous they have been severely affected by human activities, for example:

- fishing and whaling, which have upset the marine ecosystem
- the introduction of new species, such as:
  - goats, which compete with native species for the vegetation
  - dogs and cats, which chase and eat the native species

  - rats and mice, which damage the eggs of native species
  - plants such as elephant grass, which compete with native vegetation
- tourism
- disturbance due to scientific research and collection of samples
- increasing population, which requires more land for housing and agriculture, produces sewage and waste, and uses more water and energy supplies

These effects must be managed to enable survival of the indigenous species and maintain biodiversity.

The strategies include:
- searching boats and tourists for foreign species
- using natural predators to reduce pest populations
- culling feral goats and pigs
- education and fostering a culture of conservation
- captive-breeding programmes for species such as tortoises
- management of the Galápagos Marine Reserve with cooperation from local stakeholders

Antarctica

REVISED

The antarctic ecosystem is under pressure from over fishing — particularly of krill which is a keystone species. Exploitation of krill is controlled by catch size — once the catch drops to a certain level the boats must move on. Whales are protected by marine reserves where fishing is restricted. Sea birds are protected by fishing at night and only during non-breeding seasons so they are less likely to be trapped in nets.

Snowdonia

REVISED

Snowdonia is a national park with a wide diversity of rare plants and animals. It attracts walkers and climbers. Trampling damage is reduced by providing well-maintained footpaths. Draining moorland and planting conifers have reduced the value of the land as a diverse plant habitat and nesting area for rare birds. It also reduces water quality in local rivers and increases the risk of flooding. These problems have been improved by using temporary dams such as hay bales and cut branches in the drainage ditches. Farmers are also encouraged to reduce sheep grazing on the mountain itself to avoid creating bare ground open to erosion.

Farmers are being encouraged to plant hedges and take action to conserve ancient woodland.

The Lake District

REVISED

The landscape is largely a result of man's activity causing deflected succession and continued management is essential.

Farmers are offered financial incentives to reduce the use of chemicals, plant hedges and care for the diverse habitats including hay meadows, heather moorland, wetland, limestone pavement and woodland.

Management techniques include:
- planting native broadleaf species that support a wider diversity of insects and birds
- rotational timber harvesting to create a mosaic of different-aged trees

- removal of invasive species, such as rhododendron and laurel which reduce growth of other plants
- legal protection of certain habitats such as limestone pavement
- controlled burning of heathland to encourage new growth
- reintroduction of waterlogged areas by artificially raising the water table

Revision activity

Write a list of all the key terms used in this chapter. Write the meaning of the key term next to each one.

Exam tip

Remember that questions could link back to the conservation of biodiversity studied at AS and ask about the conservation techniques that are used to reduce these effects.

Exam practice

1 (a) Explain what is meant by the term 'limiting factor' when applied to a population. [2]
(b) Describe and explain the effect that removing all the predators from an area may have on the ecosystem. [3]
(c) Predators are a limiting factor. List *three* other limiting factors on the size of a population. [3]

2 (a) Describe *one* example of interspecific competition. [3]
(b) Explain how interspecific competition could lead to the extinction of one species. [3]
(c) Explain why conservationists are concerned about the introduction of goats to the Galápagos Islands. [3]

Answers and quick quiz 26 online

ONLINE

Summary

By the end of this chapter you should be able to:

- explain the significance of limiting factors in determining the carrying capacity and final size of a population
- describe predator–prey relationships and their possible effects on population sizes
- explain the terms interspecific and intraspecific competition
- describe the reasons for conservation and preservation
- explain how the management of an ecosystem can provide resources in a sustainable way, with reference to timber production and fisheries
- discuss the management of environmental resources and the effects of human activities

Now test yourself answers

Now test yourself

Chapter 2

1 $I = A \times M$, $A = \frac{I}{M}$

2 The resolution of a light microscope is not sufficient to separate objects that are closer than 200 nm. Therefore, an image magnified at greater than 1500× contains no extra detail.

3 A scanning electron microscope image is in three dimensions and a surface view. A transmission electron microscope can see details inside organelles.

4 There are 1000 nanometers (nm) in one micrometer (μm) and 1000 micrometers in one millimetre.

5 The organelles could be cut in a different plane. Organelles, and especially mitochondria, can change shape and be larger in more active cells.

Chapter 3

1 A δ^+ charge is a small charge not equivalent to a whole electron or proton charge. It is created by polarity in the molecule. Certain atoms attract electrons more strongly than others and if electrons are attracted away from one part of the molecule it leaves a δ^+ charge.

2 (a) Specific heat capacity is a measure of how much energy is needed to warm up a substance. The energy is used to make the molecules move about more. Latent heat capacity is a measure of how much energy is needed to convert a substance from one state (e.g. liquid) to another (e.g. gas). The energy is used in breaking the bonds that hold the molecules together.

(b) Water has a high specific heat capacity, which means that a lot of energy is needed to warm up a body of water. As living things are mostly water, this means that a body remains at a fairly constant temperature. Latent heat is the heat needed to evaporate water or sweat. Evaporating water off plant leaves and sweat off the skin helps to keep living things cool.

3 Amylose is a long chain of glucose. Glucose holds energy. Amylose is large and therefore insoluble, so it does not affect the water potential of the cell. The molecule is highly coiled, so it does not take up much space. Amylose can be broken down easily by hydrolysis to release the glucose units when energy is needed.

4 Cellulose is a large molecule. It has many hydroxyl groups that can form hydrogen bonds and these bind to other cellulose molecules, making fibrils.

5 Lipids are high energy compounds. The fatty acid chains can be broken down and enter respiration to release energy. Lipids are not water soluble, so they do not affect the water potential of the cell.

6 Phospholipids have a phosphate group attached to the 'head' end. This is hydrophilic and will mix with water — it will twist towards water. The fatty acid 'tails' are hydrophobic and will not mix with water — they twist away from water. A group of phospholipid molecules will orientate so that they form a globule with the heads on the outside in contact with the water and the tails on the inside acting as a barrier to the water.

7 Proteins are chains of amino acids joined by peptide bonds. A peptide bond forms between the amino end of one amino acid and the carboxylic acid end of another. Each amino acid has just one amino group and one carboxylic acid group. Therefore, it cannot form an extra bond to create a branch.

8 A polypeptide is a long chain of amino acids joined by peptide bonds. A protein may consist of one polypeptide chain or it may have several polypeptide chains. A protein may also contain a non-protein prosthetic group such as the haem group in haemoglobin.

9 Metabolically active proteins need a specific shape, which is achieved by the three-dimensional tertiary structure. This gives hormones a binding site and enzymes an active site.

10 The three-dimensional shape of globular proteins is usually very specific. If it is altered by a change in temperature, the protein becomes inactive.

11 A solution with low concentration of glucose contains less glucose than one with high concentration. If the Benedict's reagent is in excess, the low concentration of glucose will react with only some of the available Benedict's reagent, leaving some remaining. If there is a lot of glucose present, it will react with more of the Benedict's reagent, leaving less or none remaining.

Chapter 4

1 If any errors were made in the copying of the base sequence, the gene code would be changed. This is a mutation. A mutation may not produce the required protein or the protein may not function properly.

2 Reading single bases would give four codes. Reading pairs would give 16 codes. Reading triplets gives 64 codes. There must be at least 20 codes as there are 20 amino acids.

3 The DNA is a large molecule — too large to leave the nucleus. mRNA is a smaller molecule and single-stranded, so it will fit through the nuclear pores.

4 mRNA is made by matching nucleotides to the exposed bases in the DNA. It is then used as a template to line up the amino acids in the correct sequence. This can be achieved only if the bases in the mRNA are exposed so that similar unpaired and exposed bases in the tRNA can bind.

Chapter 5

1 You can imagine the need for activation energy as a boulder sitting in a hollow at the top of a hill. The boulder will not roll down the hill until it is pushed up and out of the hollow. Once over the lip of the hollow, the boulder can then roll easily down the hill. An enzyme can provide an alternative route for the reaction — removing the lip of the hollow.

2 The shape of the active site depends on the interactions between the *R* groups in the enzyme molecule. These interact through bonds such as ionic bonds between oppositely charged parts of the molecule and disulfide bonds between sulfur atoms. There are also weaker hydrogen bonds and interactions such as hydrophilic and hydrophobic interactions and forces of attraction and repulsion between charges. As a substrate molecule enters the active site, the physical presence of the molecule can interfere with the weaker interactions in the active site. Also, any charges on the substrate molecule will interact with charges in the active site and this can cause the shape of the active site to change slightly.

3 In any practical procedure involving enzymes, there may be many possible variables. Each variable has the potential to alter the activity of the enzyme and therefore change the rate of reaction. If two or more variables are changed at the same time, both may have an effect on the rate of reaction. It is not possible to distinguish between the effects that each variable has on the rate of reaction unless the experimenter can be certain that only one variable has been changed.

4 Variables such as temperature and pH need to be controlled so that they do not change during the practical test and affect the rate of reaction. A control is a separate test set up to demonstrate that a particular factor is required for the reaction to occur. It is usual to set up a control in which the enzyme, or source of enzyme, is missing to show that the enzyme is essential for the reaction to occur.

5 A non-competitive inhibitor binds to a part of the enzyme molecule away from the active site. The presence of the inhibitor modifies the pattern of bonds and interactions within the enzyme molecule. As a result, the active site changes shape and is no longer complementary to the shape of the substrate. The substrate cannot enter the active site and no enzyme substrate complexes are formed.

Chapter 6

1 Active transport (using energy from ATP) moves molecules or ions across a membrane. If the ions cannot diffuse back as quickly as they are pumped then a concentration gradient will be formed.

2 Compartments can specialise — they can keep all the molecules and enzymes associated with one process together where they are in higher concentration and are not interfered with by other processes.

3 The phospholipid bilayer stops movement of ions and polar molecules. Transport proteins (carrier proteins and channel proteins) have attachment sites for specific molecules.

4 A signal molecule may be specific to a certain target cell. It also has a specific shape. If the receptor has a specific shape that is complementary to the shape of signal molecule, the signal molecule will be able to bind only to that receptor site and will affect only that particular type of target cell.

5 The phospholipid bilayer is impermeable to charged molecules and ions. They cannot diffuse through the bilayer, so they must have an alternative route. This is through proteins that span the membrane, so it is known as facilitated diffusion.

6 Proteins are large molecules and cannot easily diffuse through the phospholipid bilayer. Also, they are too large to fit through protein pores.

7 The cellulose cell wall is strong and can withstand the pressure of water in the cell.

8 A strong salt solution has a low water potential. The cell will have a higher water potential. As water moves down the water potential gradient from inside the cell to outside, the cell loses water. Water will no longer push against the sides of the cell and it will lose turgidity.

9 The solution from outside the cell will fill the gap between the cell wall and the plasma membrane. The wall is fully permeable so the solution simply moves in through the wall.

Chapter 7

1 Mitosis produces two daughter cells that are genetically identical to the original cell. It must make a copy of the DNA so that there is one full set of genes to enter each daughter cell, otherwise the cells of a multicellular organism would not contain the same DNA and genes.
2 Cell division and growth are energy-requiring processes. Each daughter cell must have enough energy to grow. Each daughter cell must also have sufficient organelles to carry out the requirements of a living cell.
3 The end product of meiosis is four cells that are haploid. The DNA replicates before division. This means there is twice as much DNA in the cell. It must therefore divide twice to reduce the amount of DNA to half the normal amount (the haploid state).
4 Cells divide and then differentiate. Differentiation turns off some of the genes so that they are not expressed. The cell than changes shape and modifies its contents to be able to carry out its function — this is called specialisation.
5 Many cells that are all specialised in the same way form a tissue. They cannot do much on their own as they are so small. However, a large number of cells all performing the same task can achieve much more together. For example, one muscle cell could not move an organism, but many muscle cells together can generate enough force to move a limb.

Chapter 8

1 Size, surface area to volume ratio, level of activity and metabolic rate.
2 An amoeba is small, it has a large surface area to volume ratio and it can absorb all the oxygen it needs through its surface. A large tree has a small surface area to volume ratio, so its surface area is too small to supply the volume. Also, diffusion is too slow to allow oxygen to reach all parts of the body.
3 Concentration on the supply side (needs to be high), concentration on the demand side (needs to be low), thickness of barrier (needs to be thin).
4 To prevent air entering the system or leaving the system through the nose. This would alter the volume of air in the chamber and lead to inaccurate readings.
5 As the subject exhales, carbon dioxide enters the chamber. A high concentration of carbon dioxide alters the breathing rate and may become harmful. Soda lime absorbs the carbon dioxide. This causes the volume in the air chamber to decrease and allows the volume of oxygen used to be calculated.

Chapter 9

1 Double circulation ensures that the oxygenated blood is separated from the deoxygenated blood. This makes transport of oxygen more efficient. Double circulation also allows a higher pressure in the systemic side so that materials can be delivered further and more quickly. The lower pressure in the pulmonary side reduces the chance of damage to the capillaries in the lungs.
2 Artery walls are thick and contain thick layers of muscle and elastic tissue. They also contain collagen. The collagen is strong and prevents the artery wall bursting under high pressure. The inner lining of the artery wall is folded and can unfold as the pressure and volume of blood increase so that the lining does not get damaged or split. As the pressure and volume of blood increase, the wall of the artery stretches to accommodate the extra blood. Once the pressure starts to decrease, the elastic fibres recoil to reduce the diameter of the artery again.
3 Veins have much less smooth muscle and elastic tissue in their walls than arteries. They do not need to stretch and recoil as the pressure is never high. They also have much less collagen. The veins are often flattened and only become round in cross-section when they are full of blood. They also contain valves in their walls which prevent the blood flowing backwards away from the heart.
4 The volume of the ventricle chambers is the same on both sides of the heart. The wall of the left ventricle has much thicker muscle so that it can create a higher pressure to push the blood further around the body.
5 The atrioventricular valves are pushed open and closed by changes in pressure in the blood. When the atrium wall contracts and the pressure in the atrium is higher than in the ventricle, the valves are pushed open as the blood flows from higher to lower pressure. When the ventricle walls contract, the pressure in the ventricles rises above that in the atria and the valves are pushed closed as the blood tries to flow away from the high pressure area.
6 The blood flowing from the atrium to the ventricle moves more slowly than the electrical stimulus. The delay gives time for the blood to flow and fill the ventricle before the ventricle walls are stimulated to contract.
7 The pressure in the ventricle rises above the pressure in the atrium. Blood starts to move back towards the atrium, but it is trapped

in the atrioventricular valve and pushes the valve closed.

8 Air contains only about 20% oxygen. As the oxygen in the air is used up, the oxygen tension decreases and the blood in the lungs is less able to extract oxygen from the air. With less saturation of the haemoglobin, less oxygen is transported around the body, so less oxygen is supplied to the brain and respiring tissues.

Chapter 10

1 When water evaporates from the surfaces of the cells in the leaf, it creates a pull or tension on the chain of water molecules in the xylem. This suction effect reduces the pressure in the xylem vessels and they may collapse under the pressure from surrounding tissues. The lignin thickening strengthens the walls to support them against this collapse. It is important that water can continue in an unbroken chain all the way up the xylem. If a xylem vessel becomes blocked or damaged, water can pass through the pits from one vessel to another to maintain the unbroken chain. The pits allow water to pass out of the vessel into the surrounding cells and tissues.

2 Companion cells carry out the active processes that are used to load sucrose into the sieve tube elements.

3 Increasing the temperature increases evaporation, so the water potential in the air spaces of the leaf increases. This increases the water vapour potential gradient between the leaf air spaces and outside the leaf, so water vapour diffuses out of the leaf more quickly. Higher light intensity causes the stomata to open wider to allow more carbon dioxide to enter the leaf for photosynthesis. With wider stomatal openings, more water vapour can diffuse out of the leaf.

4 Transpiration is the loss of water vapour from the aerial parts of the plant. The transpiration stream replaces this loss by transporting water into the roots and up the stem to the leaves.

5 In spring when the leaf buds are opening, they require a supply of sucrose. This is used in respiration to supply the energy needed for growth and also as a building block to make new cells. The leaf is therefore a sink. Once the leaf is formed and can start to photosynthesise, it manufactures sugars that will be transported away to other parts of the plant. The leaf is then a source.

Chapter 11

1 The pathogen may remain on the outside of the vector (on its mouthparts). It will not affect the vector's tissues.

2 Phloem and xylem are transport tissues. The pathogen can be transported around the plant in the mass flow of fluids in these tissues, so blocking them reduces its spread.

3 Blood must not clot in the vessels or it could cause a blockage that may cause a heart attack or stroke. Making the process complex ensures that clots are not formed at inappropriate times.

4 All cells have proteins or glycoproteins on their cell-surface membranes. On the surface of a pathogen, these may have functions such as enabling the cell to bind to other cells or to receptor sites on the host cell membrane. The immune system can recognise these proteins or glycoproteins and use them as antigens to recognise foreign cells.

5 Injecting antibodies provides instant immunity but it is short term — the antibodies do not last long in the body. No memory cells are made, so there is no long-term immunity. If dead cells or antigens are injected, this allows the immune system to become activated and produce antibodies itself along with memory cells. These memory cells are what provide long-term immunity. However, there will be only a limited number of pathogen cells or antigens and the immune system may not be fully activated. If living cells are injected, they will reproduce and increase in number — they mimic a real infection. This activates the immune system more fully and provides complete immunity.

Chapter 12

1 Any bias in selecting sites for samples will make the results less valid. This is because the element of choice can be affected by what the sampler sees. If he or she chooses an area with fewer species, the overall measurement of diversity will be an underestimate. If he or she chooses an area with more species, the overall estimate will be an overestimate. It would also be tempting to select areas that contain rare species, which would affect the decision about how important the habitat is as an environment for rare organisms.

2 Species richness is a measure of how many different species are found in a habitat. If there are a lot of species, the habitat is more diverse and likely to be more stable. Species evenness is a measure of how many individuals there are of each species. If all the species in a habitat are represented by a large number of individuals or a large population, the population is likely to be stable. It also means that the habitat is more diverse.

3 Monitoring diversity over a number of years can reveal trends and changes in the populations of organisms. Careful monitoring with comparisons to physical factors on the site can help to gain an

understanding of the environmental factors that affect the growth and survival of organisms. If the site is likely to be affected by a development, it may be important to establish if there are rare or endangered species in the area in order to ascertain if the site is of importance for conservation.

4 Diverse habitats contain more species and are more stable and likely to survive some level of disruption. When resources are limited and only certain areas can be conserved, it is more important to conserve a habitat containing more species rather than one that contains few species or is unstable and could be damaged despite the attempts to conserve it.
5 Keeping two or more populations of a species allows independent evolution that may enhance genetic variation. Occasional cross-breeding between the populations reduces the chances of harmful genetic combinations arising through inbreeding. If one population is harmed by a disease, the second population ensures that the whole species is not wiped out. Having two populations also benefits scientific research into the species as comparisons can be made between the two populations.
6 Wild animals and plants do not respect international boundaries. If one country spends a lot of effort on conserving a species, the species does not benefit if another country allows the wholesale slaughter of that species. If trade in an endangered species is allowed in one country, this encourages poaching in the natural habitat of another country that is trying to conserve the species.

Chapter 13

1 All members of one species belong to the same genus, which in turn belongs to the same family, order, class etc. As you move up the hierarchy, the taxonomic groups get bigger.
2 Fungi have a spreading body structure like plants and are not able to move around like plants. They also have nuclei and membrane-bound organelles like plants. However, fungi do not possess chloroplasts and are not able to photosynthesise — they are not autotrophic.
3 If the DNA of two species is similar, there have been few mutations to make changes in the DNA. As mutations are spontaneous and random, there will be more mutations and more differences in the DNA between two species that have been evolving separately for a long time, i.e. they are not closely related. If there are few differences in the DNA, the species have not been separate for a long time and they are therefore closely related.
4 Genes code for the structure of proteins and for the sequence of bases in the protein. The protein contributes to the formation of a particular feature. If the base sequence in the DNA changes, the code also changes. This alters the sequence of amino acids in the protein. The new protein may not work or may contribute to the production of a different visible feature.
5 There is a change in the environment → which causes selective pressure → due to natural variation between members of the species → some individuals possess a feature that gives them an advantage → these individuals survive more easily → they reproduce successfully to pass on their genes.
6 A patient infected with *Clostridium difficile* is given antibiotics to kill the bacterium. Due to natural variation, some of the bacteria have some resistance. If the whole course of antibiotics is not completed, some of the resistant bacteria will survive. These reproduce and pass on the gene for resistance to future generations. The next generation also shows variation and some individuals are more resistant than others. Over many generations of bacterial growth and reproduction, successive generations will become increasingly more resistant to the antibiotic.

Chapter 14

1 Temperature — if temperature drops too low then molecules move more slowly, there are fewer collisions between molecules and reaction rates go down. If temperature rises too high the structure of proteins is altered due to bonds within the molecule being broken. The tertiary structure changes and alters the shape of the active site so that substrate molecules no longer fit. The enzyme is denatured and no longer functions.

   Water potential of blood — if the water potential of the blood changes this affects the water in the blood cells and in the tissue fluid. If blood water potential rises too high water enters the cells by osmosis and the cells may burst. If blood water potential falls too low water leaves the blood cells by osmosis and the concentration of all the dissolved substances inside the cells changes. The water potential of the blood affects the formation of tissue fluid and the constituents of the tissue fluid. This in turn affects the cells surrounded by the tissue fluid.

   pH — altering the pH of the blood has a direct effect on protein structure. Extreme pH (high or low) causes changes to the bonding in the active sites of enzymes and reduces their ability to function.

Blood glucose concentration — if blood glucose concentration falls too low then insufficient energy is supplied to the body cells, they are not able to respire as much and muscles get tired quickly. Too much glucose in the blood can affect the blood pressure and water potential. Glucose is lost in the urine.

Blood pressure — if blood pressure drops too low then insufficient blood reaches the extremities and tisues may be short of oxygen and glucose supplies. If blood pressure rises too high this can cause damage to blood vessels and lead to atherosclerosis. It can also damage the kidney or cause a stroke.

Blood concentration of ions such Na^+, K^+ and Ca^{2+} — the concentrations of these ions affect the water potential of the blood. If the concentration is too high the water potential drops and blood pressure rises. If the concentration is too low the water potential rises and blood pressure falls. Also there may be insufficient supply of certain ions to tissues that require them.

2 Positive feedback is usually harmful. If a small change in a parameter brings about an increased change the system becomes unstable. This leads to wild changes in the parameters and enzymes are not able to function at their optimum capacity. If positive feedback is used in a living system it can help to magnify a response quickly. However, it must only work until that response has been magnified to an appropriate level. For instance, the process of blood clotting involves a cascade effect where a few platelets releasing chemicals cause more platelets to become involved. This produces a clot quickly, to reduce blood loss at a wound. However, if this cascade were to continue unchecked all the blood in the circulatory system would eventually be clotted.

3 Heat can be gained from the environment. If the body is making too much heat the only way to lose that heat is to the environment. Since the skin is the organ that is contact with the environment it is the surface through which heat can be gained or lost.

Chapter 15

1 Carbon dioxide reacts with water to produce carbonic acid, which reduces the pH in the cell. Ammonia produces ammonium ions, which are basic and increase the pH. Changes in pH affect the structure and activity of proteins including enzymes.

2 In their cell surface membrane liver cells possess many channel proteins and receptor sites. The receptor sites are specific to adrenalin, insulin and glucagon — a different type of site for each hormone. Inside the cytoplasm are granules of glycogen to store glucose. There are many ribosomes to manufacture the wide variety of enzymes used in the cells.

3 The liver cells are very active and require oxygen for aerobic respiration — so oxygenated blood from the hepatic artery is important. The role of the liver is to treat the blood to remove impurities and toxins as well as excess substances that could be useful to the body. Therefore it receives blood from the digestive system via the hepatic portal vein. This brings in blood carrying all the substances that have been absorbed from digestion.

4 Ammonia is highly toxic, so it cannot be allowed to build up in the tissues. It would affect the action of enzymes by denaturing them. Mammals must conserve water as they only gain water through eating and drinking, so they cannot afford to dilute the ammonia a lot before excreting it. Therefore ammonia is converted to urea, which is less toxic. Fish are surrounded by water and can simply dilute the ammonia so much that it is not very toxic, and can excrete it along with a lot of water.

5 Proteins are too large to pass out of the blood through the basement membrane. Therefore they are not found in the tubule and do not need to be reabsorbed.

6 When the water potential rises above the set point the mechanism has a reversing effect to bring the water potential back to the set point. If the water potential falls the mechanism brings it back up to the set point. Any change away from the set point brings about a reversal of that change so that the water potential remains close to the set point.

Chapter 16

1 A motor neurone has a cell body in the central nervous system. The axon must carry the action potential out to the effector, which may be a muscle at the bottom of the leg. A sensory neurone has its cell body close to the central nervous system and the axon only has to carry the action potential into the central nervous system.

2 At rest the membrane is kept polarised at −60 mV inside the cell. When an action potential starts it opens sodium ion channels and positively charged sodium ions diffuse into the cell. This raises the potential from −60 mV towards 0 mV, so the membrane is getting less polarised — this is depolarisation. The process continues until the potential reaches +40 mV. When the potassium ion channels open the positively charged potassium ions diffuse out of the cell. This reduces the +40 mV charge back to −60 mV — this is repolarising the membrane.

3 When an action potential arrives at a node it opens sodium ion channels. Sodium ions diffuse through the membrane into the cell. As they enter the cell the concentration of sodium ions inside the cell increases and the area becomes more positive. This causes the ions to diffuse along the

neurone in the form of a local current. This local current carries the sodium ions along to the next node. When the sodium ions arrive at the next node their charge causes voltage-gated sodium ion channels to open, starting the action potential at the next node.

Chapter 17

1 There may be many tissues affected by one hormone — for example, adrenaline affects the muscles, the heart, the blood vessels and the lungs. The hormone needs to be transmitted all over the body to reach all these tissues. However, the hormone has a specific shape, so it only binds to specific receptors that have a complementary shape. Therefore the hormone only binds to cells that have the correct membrane bound receptor.

2 cAMP is the second messenger inside the cell. It activates enzymes inside the cell cytoplasm. The cell may be specialised to manufacture only certain enzymes. For example, in liver cells the cAMP activates phophorylase enzymes, which break down glycogen.

3 Stimulates breakdown of glycogen — releases glucose from glycogen and increases the concentration in the liver cells so that glucose enters the blood and is transported to the muscles.

Increases blood glucose concentration — the muscles have energy to contract.

Increases heart rate — increases transport of oxygen and glucose to the muscles for increased aerobic respiration.

Increases blood flow to muscles — to supply more oxygen and glucose for aerobic respiration.

Decreases blood flow to gut and skin — so that more blood can flow to the muscles.

Increases width of bronchioles to ease breathing — so that more air can enter the alveoli to increase gaseous exchange.

Increases blood pressure — to speed up blood flow.

4

Feature	Synapse	β cell
Potential difference across membrane at rest	-60/70 mV	-60/70 mV
Potassium channels at rest	Potassium channels closed	Potassium channels open
Calcium channels at rest	Calcium channels closed	Calcium channels closed
Effect of stimulation on potential difference	Depolarisation to +40 mV by opening of sodium ion channels	Depolarisation to -30 mV by closure of potassium ion channels
Effect of depolarisation	Calcium channels open	Calcium channels open
Effect of calcium influx	Vesicles of acetyl choline released	Vesicles of insulin released

5 A person with diabetes is unable to convert glucose to glycogen. Therefore there are no stores of glycogen in the liver or muscles. As a result once glucose in the blood is used up there are no stores that can be converted to raise blood glucose levels. The muscles lack energy and cannot respire as much and cannot make ATP for the muscles to use in contraction.

Chapter 18

1 A rapid movement of leaves in response to touch may help to avoid predation by scaring an insect or surprising a larger herbivore. This is seen in the sensitive plant *Mimosa pudica*. Other plants such as the venus flytrap can move to trap food. Slow response to touch can lead to the stem of a climbing plant winding around a support such as a branch of a tree. This supports the climbing plant so that it can grow into the light.

2 The two parts work in opposite ways — the sympathetic system prepares the body for activity by increasing heart rate, breathing rate and metabolic rate, along with other processes that are needed to be active, while the parasympathetic system conserves energy and allows the body to slow down.

3 The knee jerk reflex involves only two neurones and cannot be inhibited by activity in the higher parts of the brain.

4 The nervous system provides a very rapid response, which might be essential to escape danger. However, nervous stimulation may not last long because synapses can run out of neurotransmitters. The endocrine system takes a little longer to have its effect but it remains effective for longer, thus allowing a prolonged period of activity, which might be needed for flight or for fighting off an attack.

5 The electrical stimulation of the muscle runs through the membranes of the muscle. This stimulation needs to pass in all directions over

the heart to ensure coordinated contraction — the cross bridges help to carry the excitation in all directions. Also, cardiac muscle in the wall of the heart needs to squeeze the heart chamber in all directions so that the pressure inside the chamber increases. If all the fibres ran in one direction the chamber would change shape rather than reduce in volume to put pressure on the blood.

6 A muscle can contract or get shorter on its own. It is unable to stretch out again on its own. Therefore when a muscle contracts it must be stretched out by the action of another muscle. The muscles are therefore arranged in pairs that work against each other over a joint.

7 The I band is the length of a sarcomere that contains only actin filaments. As the myosin heads move to slide the actin past the myosin there is greater overlap between the filaments, so the length of the sarcomere that contains only actin gets shorter. Equally, the H zone is where there is only myosin and no actin. But as the filaments slide past one another the overlap increases, which, in turn, reduces the length of the sarcomere where there are only myosin filaments.

Chapter 19

1 White light consists of a range of wavelengths. Chlorophyll absorbs the red and blue wavelengths but reflects the green light — this is the light that enters our eye to enable vision.

2 The green chlorophyll breaks down before the accessory pigments. In autumn the leaves change colour as the accessory pigments become visible — xanthophyll is yellow.

3 Phosphorylation is the name given to the addition of a phosphate group to ADP to produce ATP. If this happens using the energy from light it is called photophosphorylation. After absorption the energy from light is held by excited electrons. These electrons pass their energy to the ATP molecule during phosphorylation. As the electron loses its energy it passes from one carrier to another — if the electron ends back at its original place in a photosystem it is known as a cyclic reaction — hence cyclic photophosphorylation. However, the electron may be passed to another molecule and does not return to its original place. This is called a non-cyclic reaction — hence non-cyclic photophosphorylation.

4 Light is used to produce ATP and reduced NADP during the light-dependent stage. If the light intensity is low then less ATP and reduced NADP are produced. These compounds are used in the light-independent stage to convert GP to TP. If less GP can be converted (because there is less ATP and reduced NADP) then GP accumulates. In order to produce glucose the GP must be converted to TP, which is used to make glucose. So if less TP is produced, less glucose is made.

5 (a) There must be enough oxygen in the water to prevent oxygen released by the plant dissolving.

(b) Carbon dioxide in the water may be limited — dissolving sodium hydrogen carbonate in the water ensures there is enough carbon dioxide to prevent it limiting the rate of photosynthesis.

(c) To allow time for the apparatus to equilibrate — so that the rate of photosynthesis is constant for those conditions.

(d) The bubbles may be difficult to count accurately — especially if they are appearing quickly. Also the bubbles may not all be the same size.

Chapter 20

1 The coenzyme A must be released to pick up another acetate molecule and feed it into the Krebs cycle.

2 Oxygen is the final electron acceptor. If electrons were not removed at the end of the electron transport chain then the chain would not transport more electrons. This means that no protons would be pumped across the inner mitochondrial membrane and no proton gradient would build up. Therefore protons would not flow through ATP synthase and less ATP would be made.

3 Anaerobic respiration does not include the Krebs cycle, oxidative phosphorylation or chemiosmosis. These processes are where most ATP is made.

4 Hydrogen atoms are removed from the substrate to reduce NAD or FAD. The energy associated with that hydrogen is used to drive the electron transport chain, the pumping of hydrogen ions and subsequently the production of ATP. As there are more C–H bonds in fats than in carbohydrates, there are more hydrogen atoms available to reduce the NAD or FAD.

Chapter 21

1 If an organism is well adapted to its environment it is making proteins that are suitable to their role. Any mutation that causes a change to the structure of that protein is likely to make it less effective in that particular environment. Therefore the mutation is harmful.

2 Energy is required to manufacture proteins. If there is no lactose in the environment, it is a waste of energy to manufacture β-galactosidase and lactose permease. The bacterium can conserve energy by not making these proteins.

Chapter 22

1 (a) A chromosome is a length of DNA combined with proteins. A chromatid is the name given to a new chromosome formed by replication up until it separates from its sister chromatid during nuclear division. After this point chromatids are called chromosomes.

(b) A dominant allele is one that is expressed in the phenotype and masks the expression of a second allele completely. A codominant allele is one that does not completely mask the second allele and both contribute to the final phenotype.

(c) A gene is a sequence of DNA bases that codes for a polypeptide. An allele is an alternative form of a gene with a slightly different base sequence.

2 Parental phenotypes: Red-eyed female; White-eyed male

Parental genotypes: X^RX^R; X^rY

F_1 gametes: (X^R); (X^r) (Y)

		Female gametes
		(X^R)
Male gametes	(X^r)	X^RX^r
	(Y)	X^RY

F_1 genotypes: X^RX^r; X^RY

F_1 phenotypes: Red-eyed female (carrier); Red-eyed male

F_2 gametes: (X^R) (X^r); (X^R) (Y)

		Female gametes	
		(X^R)	(X^r)
Male gametes	(X^R)	X^RX^R	X^RX^r
	(Y)	X^RY	X^rY

F_2 genotypes	X^RX^R	X^RX^r	X^RY	X^rY
F_2 phenotypes	Red-eyed female	Red-eyed female (carrier)	Red-eyed male	White-eyed male

Phenotypic ratio: 2 red-eyed female : 1 red-eyed male : 1 white-eyed male

3 Parental phenotypes: Red male; Roan female

Parental genotypes: RR; RW

Gametes: (R); (R) (W)

		Female gametes	
		(R)	(W)
Male gametes	(R)	RR	RW

F_1 genotype	RR	RW
F_1 phenotype	Red	Roan

No WW white individuals

4 100 – 84 = 16% show recessive characteristic. Therefore $q^2 = 0.16$ and $q = 0.4$.

$p = 1 - 0.4 = 0.6$

frequency of heterozygous individuals = $2pq = 2 \times 0.4 \times 0.6 = 0.48$

48% of the population are heterozygous.

Chapter 23

1 Each restriction endonuclease is specific to a certain sequence of nucleotides. If the same enzyme is used it will cut at the same point each time. If different enzymes are used they will cut in different parts of the DNA. When several identical long sections are cut up this will create sections of different lengths that overlap. This allows overlapping regions to be used to piece together the whole sequence.

2 If two species are closely related to a common ancestor they are not very different in evolutionary terms — so their DNA is similar. However, if two species are not closely related it means that they have evolved separately for a long time — so there have been more opportunities for mutations and their DNA is more different.

3 taq polymerase works well at 72°C — this is unusually high for an enzyme. It has a tertiary structure that is stable even at high temperatures.

4 A sticky end is a sequence of exposed nucleotide bases. This allows a complementary sequence to bind in place. It ensures that the correct piece of DNA adheres to the plasmid or chromosome vector.

5 The section of DNA placed into a plasmid contains two genes for resistance to antibiotics. One gene is complete and confers resistance to the antibiotic. However, the second gene is not complete as it has been cut open and the gene for the required substance has been inserted into it. Therefore the gene for resistance to the second antibiotic is not complete and does not confer resistance.

6 Somatic cell gene therapy affects only the cells treated. Germline gene therapy treats the fertilised egg or zygote (or all the cells in a very early embryo) so that all the cells in the body possess the modification — including the cells that produce eggs and sperm. Therefore the genetic modification is found in the sex cells and can be passed to the next generation.

Chapter 24

1 The conditions needed for the explants to grow include moisture, nutrients and warmth. These are ideal conditions for bacteria and fungi to grow — therefore, if there are any microorganisms present they will grow quickly and destroy the explants.

2 Sexual reproduction involves meiosis to produce haploid gametes. Chiasmata during prophase introduce recombinations of the alleles. Also fertilisation is random. Both these processes would change the genotype of the offspring and it would not be a clone.

3 Secondary metabolites are produced once the population is no longer growing. The population may be at a plateau or it may be declining. In continuous culture the population is maintained under ideal conditions that sustain exponential growth — therefore, no secondary metabolites are produced. In batch culture the population is allowed to grow, plateau and decline. This when secondary metabolites are produced.

4 The culture must be allowed to grow until the population is in the stable phase and then the conditions must be kept constant by adding just enough oxygen and nutrients and removing just enough of the culture and products to maintain the stable population.

5 Any trace of unwanted microorganisms must be removed so that only the required culture will grow. Unwanted microorganisms could compete for the nutrients, reducing the yield or they could damage the product by metabolising it or contaminating it with toxic by-products.

6 Initially conditions are ideal and there are no limits to growth and reproduction. However, as the population increases some factors required for growth and reproduction may become limited. For example, oxygen, certain minerals or energy supply may be limited. This reduces the rate of growth of the population so the growth curve levels out. At the plateau the population is not growing any larger and the number of births equals the number of deaths.

Chapter 25

1 If decomposition was too slow then minerals and elements would not be returned to the soil quickly. This would reduce the ability of plants to grow and the community would be less productive.

2 These plants are legumes, which have root nodules containing nitrogen-fixing bacteria. The bacteria absorb nitrogen and convert it to a form that can be used by plants to produce amino acids and proteins. As farmers harvest their crops they take away the nitrates that have been used for plant growth and it is important to replace those nitrates to maintain soil fertility.

3 During succession more and more species are able to survive in the modified conditions. Just before the climax community is reached there are many climax community species living in the area but there are also some sub-climax species left. These will be outcompeted and the diversity will decline a little as it matures to the full climax community.

Chapter 26

1 Scramble competition relies on all individuals finding what they can to eat or finding nest sites by chance. Genetic variation means that some individuals are slightly better than others at competing. These individuals will survive and breed more, passing on their alleles — so the frequency of those alleles increases. Competition by dominance hierachy means that the strongest most dominant individuals breed and pass on their alleles. If one individual becomes dominant and is the only one to breed there is likely to be less variation between individuals in the next generation.

2 Preservation simply keeps the environment as it is. If the ecosystem has already declined and biodiversity is being lost, preservation will not prevent further loss. It takes active intervention to prevent further decline and loss of biodiversity.

3 Each year a small part of the woodland is cut to ground level. Over the next few years the area cut will grow back, providing a different stage of regeneration each year. This increases the range of habitats available and the biodiversity is increased. Since a different part of the woodland is cut each year every stage of regeneration is maintained every year. This allows plants and animals to migrate round the woodland, living in the habitat that most suites them. The growth of wood from coppiced stumps is very rapid, so sustainable production can be achieved. The range of habitats created can also be used for recreational purposes.